BRAIN-MIND MACHINERY

Brain-Inspired Computing and Mind Opening

BRAIN-MIND MACHINERY

Brain-Inspired Computing and Mind Opening

Gee-Wah Ng

DSO National Laboratories, Singapore and Boston University, USA

World Scientific

NEW JERSEY • LONDON • SINGAPORE • BEIJING • SHANGHAI • HONG KONG • TAIPEI • CHENNAI

Published by

World Scientific Publishing Co. Pte. Ltd.

5 Toh Tuck Link, Singapore 596224

USA office: 27 Warren Street, Suite 401-402, Hackensack, NJ 07601

UK office: 57 Shelton Street, Covent Garden, London WC2H 9HE

Library of Congress Cataloging-in-Publication Data

Ng, G. W. (Gee Wah), 1964–

Brain-mind machinery : brain-inspired computing and mind opening / Gee-Wah Ng.

p. cm.

Includes bibliographical references and index.

ISBN-13: 978-981-279-025-5 (hardcover : alk. paper)

ISBN-10: 981-279-025-X (hardcover : alk. paper)

1. Neural networks (Computer science) 2. Artificial intelligence. I. Title.

QA76.87.N4966 2009

006.3'2--dc22

2008046699

British Library Cataloguing-in-Publication Data

A catalogue record for this book is available from the British Library.

Typeset by Stallion Press

Email: enquiries@stallionpress.com

Printed in Singapore by Mainland Press Pte Ltd

PREFACE

The grand challenge in artificial intelligence (AI)[a] is to build truly general intelligent systems,[b] systems that have the same intellectual capacity as humans, i.e. able to make meaningful decisions, learn and exhibit their intelligence in a general way in complex environments and across many different domains.

There have been many efforts made in the past decades to meet this challenge since Alan Turing, in 1950, first proposed the Turing test to validate machine intelligence. The quest has led to much progress in AI research. However, the advances made are largely in very specific or isolated domain and cannot be generalized well for wider applications. To refocus on the original vision of building a truly general intelligent system, more recently, attention has been turned towards the computational model of the human brain.

---

[a] Artificial Intelligence (AI) refers to the branch of science that attempts to create intelligence in machine. The definition here includes computational intelligence such as fuzzy, artificial neural network, probabilistic inference and evolutionary computing and intelligent agents approach.

[b] General intelligent system is used here for the same purpose and meaning as general artificial intelligence or strong artificial intelligence. In the aspect of mimicking the brain, it is also known as cognitively intelligent system or brain-inspired computing system or human-like cognitive system or cognitive computing or computational cognitive system. Note that some scientists may be more specific than others in defining and classifying these terminologies.

Unlike current AIs, which are limited in many ways, the human brains possess the ability to autonomously process information in complex environments, automatically learn relevant information, and can associate the right information for handling surprises in diverse domains and situations. The brain's processing power in terms of size, power consumption, plasticity and the robustness is something we would like an intelligent system to have.

The human brain and its intelligence continues to be a fascinating subject. With the recent advances in measuring instruments such as two-photon laser scanning microscopy and fMRI, the neuronal connectivity and circuitry of how the brain's various regions are hierarchically interconnected and organized are better understood now than ever before. Computer scientists hope that by reverse engineering of the brain, we may able to build cognitively intelligent systems that have this truly general intelligence ability.

This book provides a walkthrough on the quest for building general intelligent systems based on an understanding of the brain. It brings together diverse viewpoints and expertise from multidisciplinary communities that are interested in understanding and modeling the brain and mind machinery.

The book starts by providing an overview of how the brain is structured. Chapter 1 guides the readers through the various brain regions and their functions. Chapter 2 provides a description of the neurons and the synapse connections. It addresses how neurons communicate and how information is stored. Chapter 3 looks at the cortex architecture and describes the hierarchical structure design of the cortex. Specific hierarchical pathways in the cortex are discussed. Chapter 4 presents the different types of memory systems that researchers understand from studying the brain. Chapter 5 looks at the learning capability of our brain: How learning takes place and what are the various learning schemes. Chapter 6 addresses the issue of how emotion gives rise to cognition.

Each of these chapters (Chapters 1 to 6), the computational challenges to design and model the cortex from the regions and networks of neurons to the individual neuron's models is also discussed. The various designs of the hierarchy fashion from different stages i.e. from column or mass of cells to a few cells, and the computational design concept from simple to complex cells are presented.

From Chapter 7 through Chapter 10, the different design concepts of the whole brain from the computational to the cognitive science perceptive, are summarized. Chapter 7 and Chapter 8 summarize two unique perspectives of computing approaches to designing of the brain, namely laminar computing and probabilistic computing, respectively. Chapter 9 discusses on the higher theories of the brain and commonsense knowledge representation and generation. Chapter 10 provides the model of the entire brain based on cognitive architecture.

What are the prospects of building a truly general intelligent system? How soon can we create a machine with human-like intelligence? Chapter 11 explores these issues and the challenges. Many scientists give differing opinions. Some scientists believe that human-like intelligent systems will emerge within two decades.

Why is understanding and modeling of the brain difficult? Chapter 12 addresses the issues involved and discusses why access to today's modern instruments is still limited. With these limitations, it would be no surprise that only now and then would we hear or read about news of a new brain theory or a new discovery that gives new insight to the brain and mind. Chapter 13 presents the principles we can adopt to build intelligent systems from our current understanding of the brain. Chapter 14 concludes with a brief discussion on the theory of the mind.

The human brain remains a source of the great wonder and mystery. It exists in us and yet we do not feel its presence. We perceive with our mind day in and day out. I hope this book will provide some fresh perspective and thought in our quest to build truly general intelligent systems from an understanding of the brain.

ACKNOWLEDGEMENT

I would like to take this opportunity to thank my family, particularly my wife, Jocelyn Tho and my children, Joy Ng and Grace Ng, for their support in the course of this work. I would also like to thank my organization, DSO National Laboratories, for giving me the opportunity to take a year of sabbatical leave to do research in cognitive science area, which has led to the publication of this book; Prof. Steven Grossberg of the Cognitive and Neural System (CNS) Department, Boston University (BU), for hosting me as a visiting scholar during my sabbatical leave; and Prof. Regina Barzilay of the Computer Science and Artificial Intelligence laboratory, MIT, for co-hosting me as the affiliated faculty member.

I would like to thank the many reviewers who provided valuable input that help shape this book. First of all I would like to thank Prof. Steven Grossberg for reviewing Chapter 7 of this book on "Laminar Computing" and his guidance in a number of discussions; BU/CNS researchers' and lecturers' comments and discussions on the various chapters, particularly, Prof. Eric Schwartz for Chapter 12, Dr. Fazl Arash for part of Chapter 3, Dr. Mark Dranias for Chapters 5 and 6, Prof. Takeo Watanabe and Dr. Aaron Seitz for contributing part of Chapter 5 on 'Visual Perceptual Learning". My grateful thanks also to Dr. Rick Born, Harvard Medical School, for discussion on visual neuron's experimentation in Chapter 2; the colleges and professors in MIT's Department of Brain and

Cognitive Sciences (BCS) for their discussion on research work in cognitive science. These include Prof. James DiCarlo, Prof. Mary Potter, Prof. Joshua Tenenbaum and Prof. Pawan Sinhu with whom I have communicated; Dr. Maurizio for sharing his work on emotional state of the brain in Chapter 5; and Prof. Rosalind Picard of MIT's media laboratory, for sharing her thoughts on affective computing; and Prof. Deb Roy for his work in cognitive machines.

The following courses in BU and MIT are particularly inspiring: Prof. Eric Schwartz's Course on Computational Neuroscience (CN580/CN780); Prof. Ennio Mingolla's Course on Neural and Computational Methods of Vision (CN530/CN730); Prof. Marvin Minsky's Course on Society of Mind/The Emotion Machine; Prof. Rosalind Picard's Course on Autism and Technology; and Prof. Joshua Tenenbaum's Course on Computational Cognitive Science.

I am also most grateful to my friends and colleagues who directed me to many relevant discussions, workshops, seminars and websites. DSO colleagues working in computational cognitive systems (Dr. Teow Loo Nin, Ng Khin Hua, Tan Kheng Hwee, Chan Rui Zhong and Kelvin Goh) and the Nanyang Technological University collaborator Prof. Tan Ah Hwee for discussions and reviews.

My special thanks to Dr. Simon Overduin (MIT/BCS and BU/CNS) and Dr. Ostrovsky Yuri (MIT/BCS) who reviewed and provided input for most chapters of the book. I am most grateful to the editorial staff and artist from World Scientific Publishing who provided all the support and help in the course of my writing this book.

"It is good thing to give thanks unto the Lord..." (Psalm 92:1)

CONTENTS

Preface		v
Acknowledgement		ix
Chapter 1:	The Brain: The Center of Attraction	1
Chapter 2:	Neurons and Synapses: The Key to Memory and Learning	18
Chapter 3:	The Cortex Architecture: The Building Block of Intelligence	37
Chapter 4:	Many Faces of Memories — Investigating the Human Multiple Memory Systems	62
Chapter 5:	Learning Like a Human: How Does Learning Take Place in Our Brain?	81
Chapter 6:	Emotion and Cognition	104
Chapter 7:	Laminar Computing	129
Chapter 8:	Probabilistic Computing: The Bayesian Mind	153
Chapter 9:	Thinking Machine: Higher Theories of Brain and Commonsense Knowledge Generation	163
Chapter 10:	Modeling the Entire Brain: Biologically Inspired Cognitive Architectures	182

Chapter 11: Are We There? What Can the Computer Do Today and Tomorrow? 210

Chapter 12: Brain — A Forest Not Totally Explored: What are Some of the Issues? 222

Chapter 13: Understanding the Brain to Build Intelligent Systems 244

Chapter 14: Conclusions — The Mind That Matters 255

Glossary 263

Notes 275

References 333

Index 365

Chapter 1

THE BRAIN: THE CENTER OF ATTRACTION

"*Brain is the most complex living structure known in the universe ... The brain is what makes us human.*" Society of Neuroscience, Brain Facts 2006.

The brain has a size of about 1500 cubic meters and a surface with many folds (gyri or ridge on the cortex) and fissures (sulci or depression on the surface). It is divided into a left and a right hemisphere. The two hemispheres are connected by the corpus callosum, a large network of nerve fibers, which acts as a communication channel between the right and left hemispheres (Myers and Sperry, 1958; Sperry *et al.*, 1969). The larger part of both hemispheres is the cerebral cortex (Fig. 1.1), and this make up about 85% of the human brain's weight. In the inner part of the cerebral cortex is a system of nerve cells called the limbic system, and this connects to the brain stem and the other parts of the nervous system.

The top layer of the cerebral cortex is known as the **neocortex** ("neo" in Latin means "new", hence, neocortex means new cortex). The neocortex is about 1 mm thick (Schwartz *et al.*, 1988) and is made up of six layers. It constitutes about 90% of cerebral cortex and is commonly known as the seat of human intelligence. Humans have the largest neocortex as compared with other animals.

The average human brain has billions of nerve cells or neurons. Most literature gives this figure as approximately 100 billion. The latest finding from Rabinowicz and her collegues indicates that the average human brain has about 23 billion neurons (Rabinowicz *et al.*, 2002). Each of the neurons has about 1000 to 10 000 synapses, and has roughly 100 million meters of wiring. All this is packed into our brain, which if unfolded, will give a structure the size and thickness of what is like a formal dinner napkin (Webster's Medical Dictionary).

The whole human body has about 100 watts of energy and the brain takes up about one fifth of this energy, i.e. about 20 watts. Assuming 50% of this energy is dissipated, that leaves the brain with 10 watts of energy. With 10 watts and about 23 billion cells, the brain is able to function extremely efficiently and effectively in processing information – and this is what makes us intelligent beings. How does the brain do this and what are the mechanisms involved?

This chapter begins by presenting the wonder of the brain's organization, its early state of innate regionalization, and specialization during the development and growing process. How does the brain wire itself, and what are the estimated computational power and capacity of the brain?

How is the Brain Organized? Specificity Implies Regionalization and Specialization

Genes appear to "encapsulate a program", which directs and specifies how and where the early nerve fibers (or axons) grow, are connected, and how the migration of neurons from their birthplace to a final destination should take place in the development of the brain. The first set of tracts in early brain development forms the basic "scaffolds" on which most later tracts form (Easter *et al.*, 1993; Wilson *et al.*, 1990). The specific tract formation or basis of cellular structure suggests the early brain regionalization and development of specific functional areas (Moore *et al.*, 1996). The different regions of the brain and their functions are largely mapped.

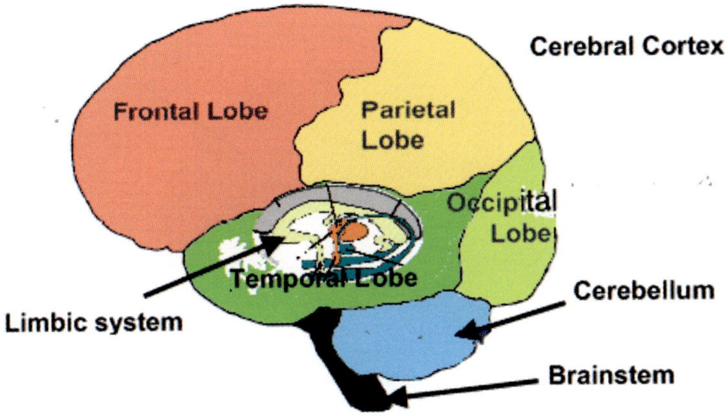

Fig. 1.1. The four major parts of the brain, namely, the cerebral cortex, cerebellum, limbic system, and brain stem.

The cerebral cortex consists of four lobes or regions, namely, the frontal lobe, parietal lobe, temporal lobe, and occipital lobe (Fig. 1.1). The cerebral cortex occupies the largest part of the human brain, followed by the cerebellum.

The brain stem (Fig. 1.2) is basically made up of the midbrain, pons, and medulla. The brain stem is the lower extension of the brain where it connects to the spinal cord and the rest of the nervous system. It controls vital life functions such as breathing, heartbeat, and blood pressure.

The corpus callosum connects the right and left cerebral hemispheres of the brain and is made of millions of nerve fibers. The fibers are the axons of the neurons from the cerebral cortex. The thalamus is the center of the brain, where it connects to the rest of the nervous system via the brain stem.

The two hemispheres (left and right) are each divided into four regions or lobes (Fig. 1.1). Roughly speaking, the four lobes perform the following role:

- *Frontal lobe* (behind the forehead): is associated with movement control and high-level cognitive tasks, such as speaking (the control of many muscles used for speech), planning,

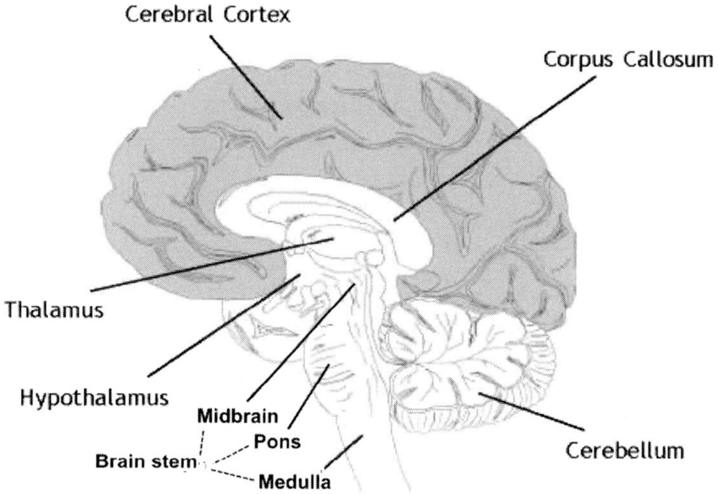

Fig. 1.2. The inner structure of the brain. (Picture adapted from http://www.ohsu.edu/thinkfirst/brain.html.)

problem solving, reasoning, the ability to make judgments, and moral behavior.

- *Parietal lobe* (top and rear): is associated with the processing of somatosensory information, i.e. sensing and perception of sensory stimuli from the skin (touch information) and internal organs, as well as mathematical and spatial reasoning.
- *Occipital lobe* (back): is largely for the processing of visual information. The primary visual cortex is at the occipital lobe.
- *Temporal lobe* (just above ears): is associated with the processing of auditory information (at the primary auditory cortex), speech perception, and object recognition. Various parts in the temporal lobe such as the hippocampus are associated with aspects of memory and spatial navigation; the amygdala is associated with emotional and affective behavior.

Many theories have sprung up that explain the differences between left and right hemisphere functioning. For example, according to some theories, the right hemisphere is more closely associated with perceptual tasks (object recognition such as face, picture, etc), processing of spatial information, parallel processing, creativity, and emotions. Pilots are said to be more right-brained people. In contrast, the left hemisphere is closely associated with speech, reading, language (including sign language), arithmetic, calculation, logic, sequential processing, and the ability to analyze. Writers are said to be more left-brained people.

Increasing research on split-brain[a] patients shows that the two hemispheres have different styles of information processing. It is noted that the left is more bias towards details, while the right leans more towards a holistic outlook. There is also some evidence that the right hemisphere is more specialized for negative emotions, and the left, more for positive ones.

Figure 1.3 shows a more detailed diagram of the brain's various parts. The spinal cord connects the brain stem (the medulla is part of the brain stem) to the rest of the body's nervous system. Further up the brain stem is the limbic system, which consists of the thalamus, hypothalamus, amygdala, olfactory bulb, hippocampus, and basal ganglia. The limbic system, which plays a vital role in human emotional responses, is linked to the cerebral cortex. More details of the limbic system diagram can be found in Chapter 6. The major part of the cerebral cortex consists of the visual cortex (which receives information from the eye via lateral geniculate nucleus (LGN) to the primary visual cortex (V1)), the auditory cortex (which receives information from the ear), the somatosensory cortex (concerned with receiving general sensations); the motor cortex, and the prefrontal cortex. The prefrontal cortex is important for higher cognitive function such as mental processes for reasoning, problem solving, and decision-making. The motor cortex, which includes the primary motor cortex (M1) and secondary cortices, are importance for the planning, control, and execution of

[a] This is a term to describe the result when the corpus callosum connecting the two halves of the brain is severed to some degree.

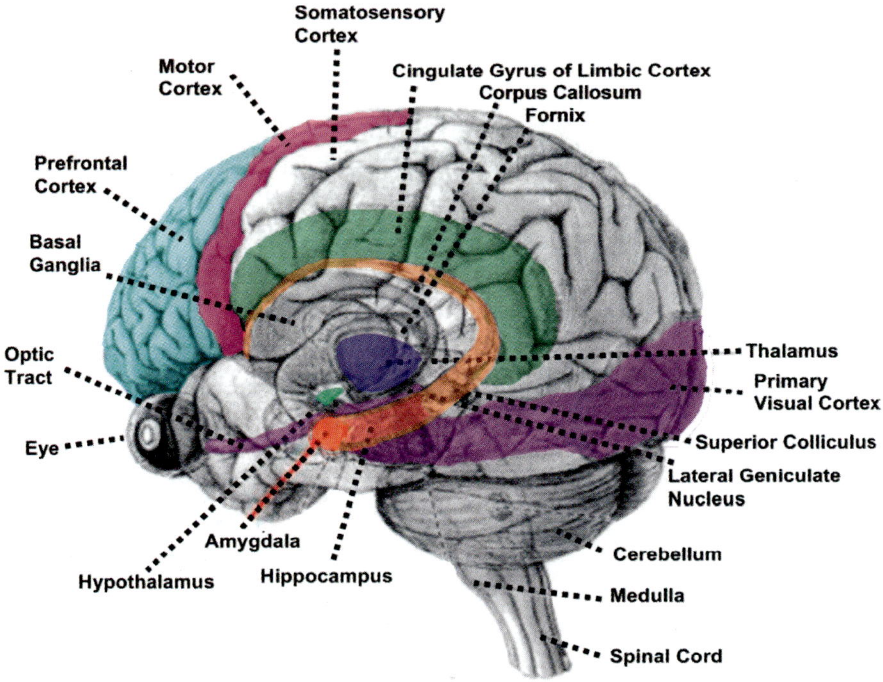

Fig. 1.3. A detailed diagram of the brain's various parts.

voluntary motor functions. Note that the diagram and the various areas are not drawn to proportion.

Visual processing constitutes at least 50% of the brain processing and is one of the critical functions of the brain. Figure 1.4 shows the visual processing pathway from the retina through the lateral geniculate nucleus (LGN) to the visual cortex via the optic nerve. The visual cortex includes the primary visual cortex (also known as the striate cortex, V1) and the extrastriate cortices (V2, V3, V4, and V5).

The first layer of the visual cortex is V1 the striate cortex (or primary visual cortex). The subsequent layers include V2, V3, V4, and V5. The occipital lobes cover V1 to V5. The visual area V5 is commonly known as MT (middle temporal), and it plays a major role in the perception of movement.

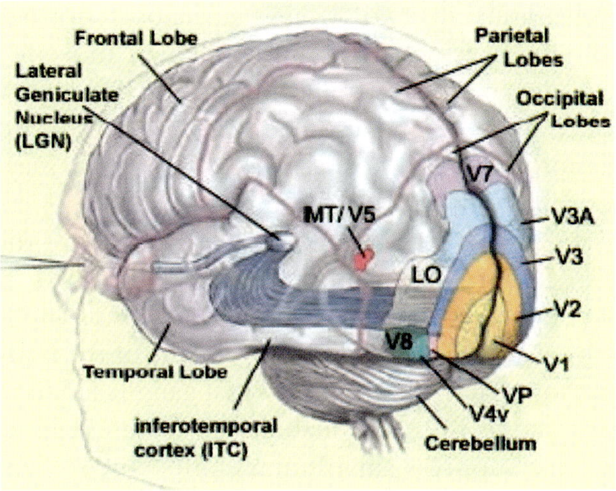

Fig. 1.4. Diagram showing the visual pathway from the retina to the visual cortex via the optic nerve. (Picture from http://thebrain.mcgill.ca)

The visual cortex has two functionally specialized processing pathways, namely, the occipitotemporal pathway (also known as the ventral stream or "what" stream) and the occipitopariental pathway (also known as the dorsal stream or "where" stream). The occipitotemporal pathway is crucial for object identification while the occipitopariental pathway is crucial for appreciating the spatial relationships among objects (Ungerleider and Haxby, 1994). Hence, a patient who has a lesion on the occipitopariental pathway (but with the occipitotemporal pathway functioning normally), if presented with a pencil horizontally, will recognize that it is a pencil, but will not be able to describe the orientation or grasp the pencil in the correct orientation.

Other examples of specific functional areas are:

- The hippocampus (Fig. 1.3), located on the temporal lobe of the brain, plays an important role in the formation of declarative memory and spatial navigation. Hence, if one has a lesion on the hippocampus, one may not be able to remember a new event that has just happened a few minutes ago. He may annoy everyone by forgetting what was discussed just a moment ago.

- The cerebellum (Fig. 1.3), located at the posterior bottom part of the brain, is essential to the control of movement of the human body. It acts as a reflex center for the precise maintenance of equilibrium and also regulates and coordinates posture and all motor activity. Losing this brain region may result in poor coordination of body movement.
- The basal ganglia (Fig. 1.3), located at the base of the cerebral cortex, are important for movement selection. The basal ganglia are interconnected with the cortex and thalamus and are associated with muscle-driven motor movements of the body. Clinical data has shown that lesions in the basal ganglia led to movement disorder ranging from the inability to initiate a movement to inability to suppress involuntary movements and causes Parkinson's and Huntington's diseases.

  The basal ganglia and cerebellum are large collections of nuclei that modify movement on a minute-to-minute basis. The motor cortex (Fig. 1.3) sends information to both the basal ganglia and cerebellum, and both structures send information right back to the cortex via the thalamus. (Remember, to get to the cortex, you must go through the thalamus.) The output of the cerebellum is excitatory, while that of the basal ganglia is inhibitory. The balance between these two systems allows for smooth, coordinated movements, and a disturbance in either system will show up as movement disorders.
- The amygdale (Fig. 1.3), located deep within the temporal lobes, is involved in processing human emotion. Experiments have shown that lesions of the amygdala in humans affect emotion, such as the abolishment of fear conditioning (Armony, 2007). Hence, if a patient has the amygdale damaged, he may not be fearful if a poisonous snake is placed in front of him.
- The thalamus (Fig. 1.3), located at the top of the brain stem, acts both as the central control, where many functions are linked, and as a relay to the cortex, i.e. it relays information to and from the cortex. Many sensory functions are linked to the thalamus, and hence, one cannot imagine what will happen if the thalamus is damaged.

- The superior colliculus (Fig. 1.3), located just below the thalamus, is responsible for the generation of saccadic eye movements and eye-head coordination.

From the broad study of the brain, in some general sense, we can conclude that the brain is divided into functional regions. Each region performs specific tasks and is specialized; this indicates the specific **spatial representation** of the brain. There is strong **association and interconnection between the different regions** to make the system "operate" as a whole.

How localized is the function in a region is a debatable issue. Some functions are found to be pretty localized, e.g. the fusiform face area (FFA) in the human extrastriate cortex is found to be critical for human face processing, and is relatively localized and very specific for human face recognition (Kanwisher *et al.*, 1997). The parahippocampal place area (PPA) and retrosplenial cortex (RSC) are also found to be more specialized for the identification of objects (such as a house), spatial structure, and location in a visual scene. (Epstein *et al.*, 2003; Epstein and Higgins, 2006). Hence, if a patient has a lesion on the FFA (Fig. 1.5), he may not be able to recognize faces, such as his wife or mother, and if his PPA and RSC are damaged, he may not be able to recognize his way home. However, there are debates on how locally specific these areas[b] are. For example, (Haxby *et al.*, 2000) Haxby and colleagues reported that the FFA may not be as localized as it seems. However, more recent studies have increasingly supported the hypothesis that these are very localized and specific functional areas.

Nevertheless, most scientists' believe that many functions remain more distributed than localized. "Distributed" here refers to a set of neurons (currently I do not have good reference to support how many thousands of neurons) are involved either within

---

[b] For instance, while the FFA may be inevitably active in facial perception, the function of facial perception clearly involves activation of many cortical and sub-cortical areas, initially driven in a sequential fashion by visual input.

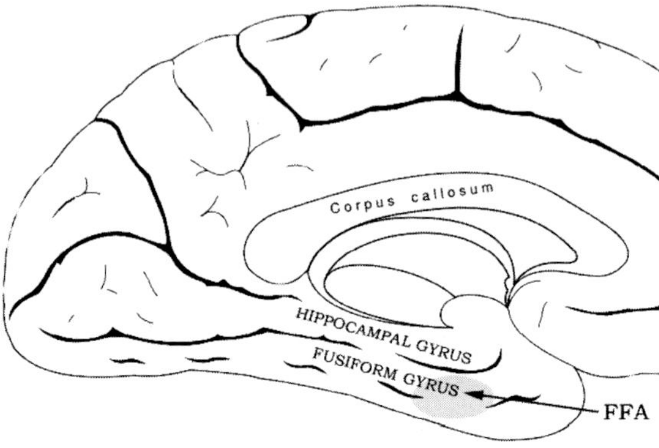

Fig. 1.5. Diagram showing the hippocampal gyrus and the fusiform gyrus, where the parahippocampal place area and fusiform face area are located respectively.

and/or across the regions and areas in representing a specific object or concept. This is to say that the functions are mediated by distributed neural systems in the brain. The theory that supports this is known as the "**Distributed Representation Theory**". However, how extensive the distribution is within the regions and across the multiple regions remains fuzzy for some functions. On the other hand, the "**Grandmother Cell Theory**" is an extreme case of supporting a very specialized neuron that represent a specific concept or object. For example, in the "Grandmother Cell Theory", a single specific neuron is used to recognize the face of my grandmother, and in the "Distributed Representation Theory", a distributed group of neurons are needed to recognize the face of my grandmother. Both the "Distributed Representation Theory" and "Grandmother Cell Theory" are further discussed in Chapter 2.

Other interesting research issues on regionalization and specialization in the brain include:

- How does each region in the brain actually work and develop?
- Why are some essential functions such as facial recognition very specifically localized in the brain while others are not?
- To what extent can the specialized regions reorganize in the event of injury to the brain?

- How interdependent are some of the specialized regions with other regions in the brain?
- Which of these regions are uniquely human and which can also be found in other primates?

How Does the Brain Wire Itself?

MIT Professor Sur and his team have tested and demonstrated that the visual signals can be rewired to the auditory cortex. They demonstrated these using ferrets in their early development stage of life. The result shows that orientation selectivity appeared in the rewired auditory cortex, statistically identical to the visual cortex. The ferrets had been successfully trained to perform vision tasks using the auditory cortex (Sharma *et al.*, 2000).

Biologists have shown that a baby ferret's brain is one of the least developed among other animals during birth (i.e. ferret brains are immature at birth and undergo a late form of development). Hence, Sur's team chose baby ferrets to demonstrate that visual signals can be rewired to the auditory cortex. This experimental

Fig. 1.6. A picture of the ferret.

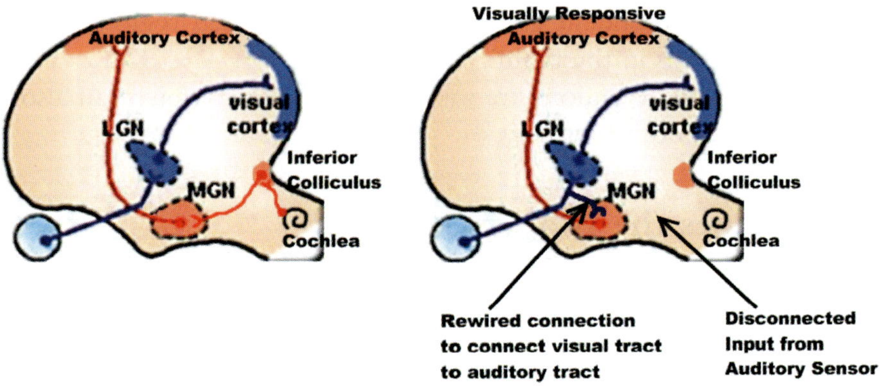

Fig. 1.7. Left picture shows a normal wired cortex; Right picture shows a rewired cortex. (Diagrams modified from paper (Sur, 2001))

discovery was highlighted several times in the news in 2000 (see also http://web.mit.edu/newsoffice/2000/neuromit.html). Later experiments also suggest that the brain is so adaptable that all the five senses can be rewired (Abrams, 2003). This adaptability of the brain is also known as plasticity.

Sur's latest research finding indicates that although a rewired auditory cortex takes on many properties of the visual cortex, it retains some of the original auditory network properties (Majewska and Sur, 2006). Rewiring appears not to alter the structure of thalamocortical arbors and the intracotical connection. Majewska and Sur indicated that the rewired auditory cortex is not as regular and precise as the visual cortex. Orientation maps in a rewired auditory cortex show less periodicity and larger orientation domains than those in the visual cortex. Sur in his article also added that on a synaptic level, spine dynamics in a rewired auditory cortex are more similar to those observed in a normal auditory cortex than those in the visual cortex, where spines are more stable. (Spine dynamics means the change in density of the dendritic spine. See notes on Chapter 1 for details.) These findings argue that although activity has a powerful ability to remodel cortical connection, it probably acts in concert with a scaffold of connectivity laid down by intrinsic cues (intrinsic cues mean stimulus or actions coming from in-born (nature) or built-in).

The ability of the brain to rewire itself suggests that:

- Biologically **neural cortical columns** in different sensory are, in general, functionally similar in their neural cell type and laminar structure.
- There exist **adaptive or learning mechanisms** across the cortex. The ability to learn is due to the plasticity of the brain. The brain is plastic in the sense that it can reorganize the neural pathway based on new stimuli or experiences.

With exposure, experiences, or activities from different modalities (sensory systems), the neural cortical columns learn and develop different functional mechanisms and properties.

However, the brain in early development starts to partition and form specific pathways for specific functions and regions. Although the **pathways can be rewired** for other modalities input, it is not as efficient due to the early stage of the pathway development and properties. **Genes** probably dictate the overall framework or architecture of the brain and the basis columar organization structure of the neocortex. The details of the columar organization are "filled-in" by the environment or external stimuli. To illustrate further, genes are the chief architects that decide all the regions and form the main connections of the major pathways. This is analogous to a building in which the chief architect decides all the main structures and compartments of the building. The details are filled-in by the tenants who will occupy the building. In the case of a building designed for, say, shopping purposes but is then put to residential use, it will still work but may not be optimal in its usage.

What is the "computational power" of the brain?

There are several approaches in estimating the "computational power" of the brain. Before we proceed further, I would like to note that using the term "computational power" to describe the brain's capability of processing information may not be the best approach, since the brain is not entirely "a computing machine" *per se* and does not crunch numbers. However, due to the lack of a better terminology and also as the main quest of the brain's modeling technique in this discussion uses the computer, we will use the computer science terminology of "computational power" to measure the brain with respect to operation cycles per second.

Three possible approaches in measuring the brain's "computational power" are:

- The first approach is to count the number of synapses, guess their speed of operation, and determine the number of synapse operations per second. There are roughly 10^{15} synapses operating at about 10 impulses/second, giving roughly 10^{16} synapse operations per second (Kandel, 1985).
- The second approach is to estimate the computational power of the retina and multiply it by their brain-to-retina size ratio. Assuming that a typical ganglion cell in the retina performs about 100 analog additions 100 times per second, then computation of the axonal output of each ganglion cell requires about 10^4 analog additions per second. Given that there are about 10^6 axons in the optic nerve, a total of 10^{10} analog additions per second is arrived at. There are about 10^8 optic nerve cells in the retina and between 10^{10} and 10^{12} nerve cells in the brain, so the brain is roughly 100 to 10 000 times larger than the retina. Using this approach, the brain is estimated to be able to perform about 10^{12} to 10^{14} operations per second (Moravec, 1998b).
- The third approach is to divide the total useful energy used by the brain per second by the amount of energy used for each basic operation to give the maximum number of operations per second. A brain has about 20 watts. Given that about half of this energy is lost through dissipation, the effective utilization is about 10 watt of energy. Using 10 watts as calculation, we can perform about 10^{15} operations per second (Merkle, 1989).

From these three approaches, it is concluded that the estimated computational power of the human brain is about 10^{13} to 10^{16}

operations per second or 10 million to 10 000 million MIPS (this is about 10 teraflops[c] to 10 petaflops). Note: there are about 10^{11}–10^{12} neurons in the brain communicating with each other over about 10^{14}–10^{15} synapses. If the brain capacity is measured by the number of synapses, then 10^{14} synapses will be 100 terabytes. This is assuming 1 synapse = 1 byte.

Using this information (operations per second/memory capacity), many scientists then predict how the computer will match the human brain. For example:

> "*At the present rate, computers suitable for humanlike robots will appear in the 2020s.*" (Moravec, 1998a,b)

> "*The memory capacity of the human brain is about 100 trillion synapse strengths, which we can estimate at about a million billion bits. ...Five billion bits per $1k in 1998 will double 17 times by 2023, which is about a million billion bits for $1k in 2023.*" (Kurzweil, 1999)

> "*It is estimated that the storage size of a human adult brain is approximately 10^{14} bytes. .. If the disk price continues to drop at this rate, then the hard disk of a human brain size could be purchased with $6000 by 2008; i.e. many desktop computers could afford human brain size storage at that time.*" (Weng, Aug 2006)

Assuming the computation power of the computer continues to accelerate (as per Moore's law), and within economic scales, we arrive at a projection based on the current growth trend in computational power of computer: within 20 years, we will have sufficient computational power, at a relatively affordable cost, to perform the number of operations **(about 10 teraflops to 10 petaflops)** that is in some sense "equivalent" to the brain. Note that the current

---

[c] Flops is an acronym meaning floating point operations per second. One teraflop is equal to about 10^{12} flops and one petaflop is equal to about 10^{15} flops.

supercomputers, such as the IBM BlueGene/L (performance at 259 teraflops — winner of the HPC challenges in November 2006) and MDGRAPE-3 (developed by the RIKEN research institute and cited in their institute's report, dated June 2006, to perform at 1 petaflops), can do the job. The IBM Blue Brain project used BlueGene to do research on modeling the neocortex and attempted to prove that the brain can be simulated at the cellular level. The IBM Blue Brain project started in 2005.

The storage capacity of the brain, if based on **100 terabytes**, is already achievable. Hence, the more important issue is not speed or capacity, but how (much) we can model after the brain to build cognitively intelligent systems.

Summary

The brain is organized into regions, and each of the regions has a specialized function: Each of these regions is heavily related to one another and scientists have not fully understood how the different regions are interconnected. Genes have a strong influence on the fundamental design of the columnar organization and structure of the brain. The details of the columnar organization and structure are "filled-in" by the environment or external stimuli. In this way, the brain is very adaptable to its environment and can be nurtured. Genes also influence the many specific pathways and their properties. This fixed wiring influenced by genes forms the nature[d] here refers to the natural aspects versus the nature aspect part of the brain. The brain machinery gives rise to the mind and forms the basis of our behaviors.

In our quest to reverse engineer the brain to build intelligent systems, it would be important to understand:

- How the many neurons or cells store information.
- What the essential regions and their functions are.
- How the regions are structurally organized.

[d] Nature here refers to the natural aspect versus the nurture aspect.

- How the regions are interconnected and associated with one another.
- What the essential pathways, network and their properties are.
- How the different regions in the brain orchestrate together as a complete system.

Chapter **2**

NEURONS AND SYNAPSES: THE KEY TO MEMORY AND LEARNING

"*...memories might be stored in the connections between nerve cells.*"
Santiago Ramon y Cajal's 1894 lecture.

In 1894, Santiago Ramon y Cajal (Fig. 2.1) first predicted that memories might be stored in the synapses. The synapse is the nerve cell's junction where cells are interconnected. Today, the synapse is understood to be the key basis for learning and memory.

Fig. 2.1. Santiago Ramon y Cajal (1852–1934). (Picture from http://nobelprize.org/nobel_prizes/medicine/laureates/1906/)

Fig. 2.2. Heinrich Wilhelm Gottfried von Waldeyer Hartz (1836–1921). (Picture from http://en.wikipedia.org/wiki/Heinrich_Wilhelm_Gottfried_von_Waldeyer-Hartz)

In 1906, Santiago Ramon y Cajal and Camillo Golgi won the Nobel Prize together, in recognition of their work on the structure of the nervous system.

A few years earlier, Heinrich Wilhelm Gottfried von Waldeyer Hartz (Fig. 2.2), a German anatomist, coined the term "neuron" to describe the nerve cell. In 1891, Waldeyer Hartx was honored in the neurosciences as the founder of the so-called "neuron theory" to describe the basic structural unit of the nervous system.

The brain is a massive neural network of neurons (or nerve cells) that communicate with one another via synapses (cell junctions or nerve cell connections at the end of the terminal button, which is a junction gap). Synapses are viewed as so essential in the make-up of our brain that one even quoted: "*You are your synapses. They are who you are.*" Joseph LeDoux, 2002 (in *Synaptic Self*).

As you read this chapter, your synapses should change as new information is stimulating you. In 2008, the word "synapse" turned 111 years old. The word synapse was first used in a book called, *A Textbook of Physiology, part three: The Central Nervous System*, by Michael Foster and assisted by Charles S. Sherrington, in 1897.

It is commonly cited that Charles S. Sherrington coined the term "synapse". The word "synapse" comes from the Greek word, meaning "to clasp together".

Fig. 2.3. Charles S. Sherrington (1857–1952). (Picture from http://nobelprize.org/nobel_prizes/medicine/laureates/1932/sherrington-bio.html.)

This chapter presents the following:

- How do neurons communicate and affect our learning and memory?
  - Neurons have specialized functions.
  - How is information transmitted and stored?
  - How is information represented?
  - How does the neuron's receptive field affect our visual scene?
- How does synaptic plasticity give rise to learning and memory?
  - Synaptic plasticity.
  - Synaptic spike and normalization.
- Modeling the neurons and synapses
  - Historical background of artificial neural networks.
  - Modeling at the cellular level.

How Do Neurons Communicate and Affect Our Learning and Memory?

The fundamental building blocks of the nervous system are the neurons.

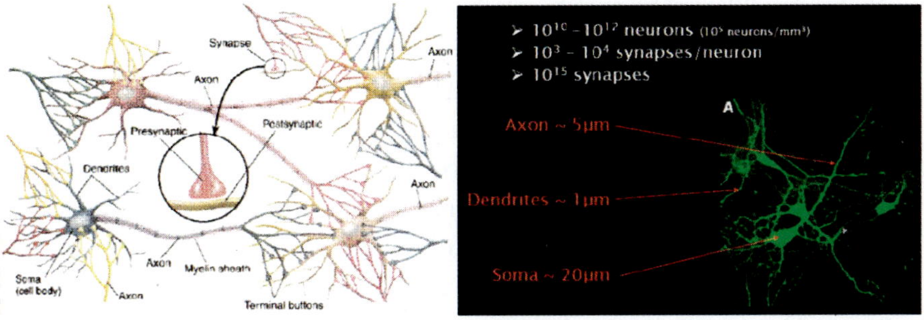

Fig. 2.4. Pictures showing the neurons, dendrites, soma, axon, and synapses. (Picture from MIT/BCS)

The neuron (Fig. 2.4) consists of a cell body called the soma, which contains the nucleus for sustaining the life of the cell, dendrites that serve to receive information from other neurons (via synapses), and an axon that carries information away from the soma to the terminal buttons where the information is transmitted to other neurons. At the end of the terminal buttons are gaps known as synapses. Neurotransmitters are secreted from the terminal buttons. Note that neurotransmitters are specialized chemicals secreted from terminal buttons; there are more than 30 different neurotransmitters. These neurotransmitters pass into specific receptor sites on the dendrite and enter the cell body.

The axon, also known as the fiber, is surrounded by fatty tissues, which serve as an insulator. These fatty tissues increase the speed of neural impulse transmission. The neural impulse transmission speed varies, but in general it is in the range of 2 to 200 mph or of the order of milliseconds (1000th of a second). Note that the current speed of computers of a single pulse is of the order of nanoseconds (one billionth of a second).

Neurons have specialized functions. Research on human and animal visual systems shows that the visual cortex has many **specialized neurons**.[a] For example, some neurons respond to edges,

[a] Note that the specialized neurons may be arising due to their connections in the network and the stimuli they are exposed to early in life.

others for filling in, and yet others respond only to certain directions, while some others are sensitive to color contrast, and bright or dark patterns. By pooling all the different responses from the neuron population, the brain is able to recognize objects, see shapes, and compare the balance between different degrees of bright and dark patterns. To illustrate this further the following illusions are shown (Figs. 2.5, 2.6 and 2.7).

Figure 2.5 shows that the straight edges of pac-man features appear to extend beyond their endpoints by extrapolation, where

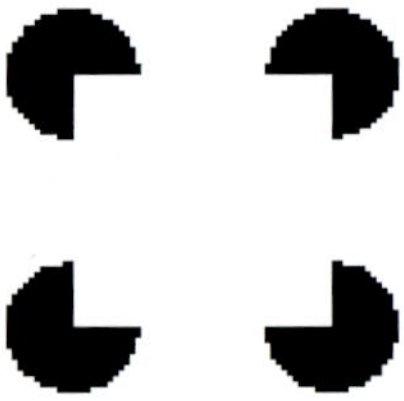

Fig. 2.5. A white square illusion contour formed by four pac-mans. The white square is formed due to edge extension and the filling-in effect of our visual process.

Fig. 2.6. The illusion of objects rotating. We perceive the rotation due to the firing up of the visual cortex neurons responsible for specific directions of motion when stimulated by the different intensities of colors and shapes of the picture.

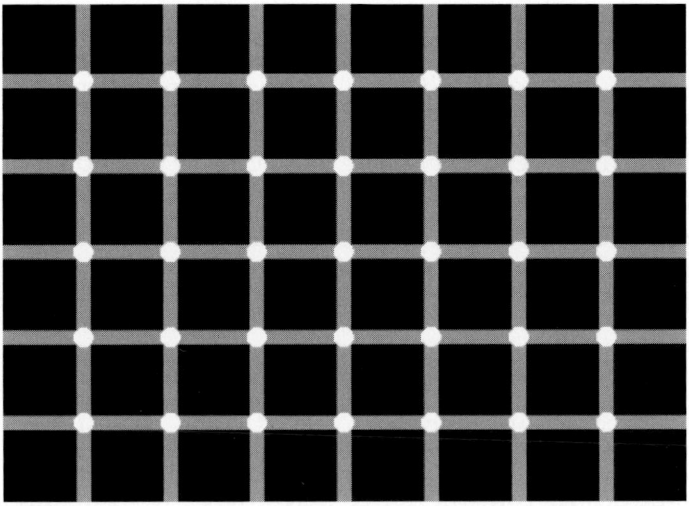

Fig. 2.7. The illusion of color contrast. (Picture from http://www.eyetricks.com/illusions.htm.) Can you count the number of black dots? Black dots are seen due to inhibition and excitation of neurons responsible for receiving light.

they meet with other extended edges coming from the opposite direction to form the illusory contour. The four contours in turn suggest the whole square as an enclosed figure. The contrast of this white illusory square against the black circles appears to be brighter than the white background against which it is seen, and this perceived brightness is observed to pervade every point across the whole illusory surface uniformly. (Grossberg and Mingolla, 1985) have modeled this perceptual computation as a spatial diffusion of color experience from the edges across which the dark/light contrast signal originates, filling in the entire illusory surface.

Figure 2.6 shows the illusion where the human eye perceives that the objects are rotating. The effect is due to the fact that in the visual cortex, there are neurons that respond to specific directions of motion. Why are these neurons fired up? One likely cause is that the different intensities of colors and shapes at the different regions are equiluminous, and this causes a mismatch in the correspondence periphery, which causes the motion neurons to respond and yield a perception of motion. Hence, the motion neurons are fired up and the brain "sees" the objects "rotating". Another example is when a person looks at an object moving downward, such as a waterfall, and that person's downward receptors (sensory neuron that responds to this scene) are in action. If he stares at the downward motion long enough for those cells to become fatigued and then looks at a stationary object, like a grassy hill, the grass will appear to be moving upwards. The upward receptors compensate for the fatigue of the downward receptors. This phenomenon is known as an after affect.

Figure 2.7 illustrates how contrast affects color perception. This is known as Hermann's Grid. Can you count the number of black dots? The areas at the corners of the black boxes appear gray. This happens because of something called lateral inhibition. In the retina, when some light-receiving cells are activated others around them shut down. You will notice that where the white lines intersect, there is black on four sides, whereas the lines themselves are surrounded by black on only two sides. When you look at the intersections, the cells in the retina are surrounded on four sides by other cells that are also receiving light. They are therefore more inhibited than the cells focused on the lines. It is their inhibition that causes the dark spots to appear.

How is information transmitted and stored? Synapses allow neurons to communicate with one another through axons and dendrites. There are generally two forms of synapses, namely, electrical and chemical. Electrical synapses provide for rapid transmission of information as electric current passes from one neuron to the next. In contrast, chemical synapses involve the transmission of electrical signals arriving at the axon terminal of one neuron via the release of a chemical messenger such as a neurotransmitter or neuromodulator.

Figure 2.8 shows how information flows between two neurons via the synapse. The two interacting neurons are the presynaptic neuron from the axon terminal and the postsynaptic neuron from the dendrite spine. Action potential is the electrical signals that travel down the axon carrying information from the neuron's cell body to the axon terminal. Action potential is generated by ions, which are charged particles. The most common ions are potassium (K^+),

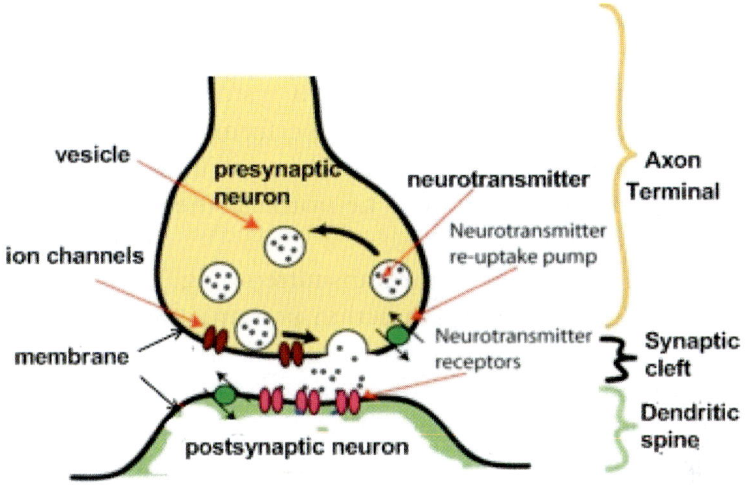

Fig. 2.8. The interaction between the presynaptic neuron and postsynaptic neuron.

sodium (N^+), calcium (Ca^{2+}), and chloride (Cl^-). Ions cross the membrane of a neuron through the ion channels. However, ion channels will only open for ions to flow through when a specific neurotransmitter binds with the receptor. Once the neurotransmitter binds with the receptor, the flow of ions through the open ion channels produces small electrical changes in the membrane of the postsynaptic neuron. These small electrical signals are called synaptic potentials (or sometimes also known as synaptic strength). The postsynaptic neuron will fire an action potential (sometime known as an excitatory effect) when and only when the sum of all its synaptic potentials equals or exceeds its threshold. The sum of the synaptic potentials is typically contributed by many neurons as each neuron typically forms synapses with many other neurons.

On the other hand, a presynaptic neuron may release a neurotransmitter that inhibits or reduces the synaptic potential in the postsynaptic neuron, i.e. decreasing its excitability, and therefore, decreasing the neuron's likelihood to fire an action potential. Excitatory (promoting an action potential) and inhibitory (reducing the likelihood of an action potential) effects are due to the type of neurotransmitter released, as well as the receptor at the postsynaptic neuron. The main inhibitory neurotransmitter, GABA, has an inhibitory effect in mammalian systems. Note that nearly 80% of the neurons in the brain are performing excitatory effects and the remaining 20% are involved in inhibitory activities, i.e. there are many more excitatory neurotransmitters than inhibitory neurotransmitters.

The vesicles store the neurotransmitters (Fig. 2.8). Vesicles are produced in the cell body. An action potential causes a vesicle to migrate to the membrane, bind to it, and then release its neurotransmitter contents to the postsynaptic neurons. Neurotransmitters are chemical substances released from a neuron at a synapse, and are used to relay, amplify, and modulate electrical signals between a neuron and other neurons. A neurotransmitter conveys information between two neurons, while neuromodulators convey information to a region of neurons (or group of neurons). There are many different types of neurotransmitters and neuromodulators, corresponding with different types of receptors. A receptor is a protein on the cell membrane that binds to a specific molecule, such as a neurotransmitter, hormone, or other substance, and initiates the cellular response of that specific molecule. More information of the different types of neurotransmitters, neuromodulators, and receptors can be found in (Alexander and Peters, 2000) or at SenseLab (http://senselab.med.yale.edu/). The SenseLab project is managed by Professor Good Shepherd from Yale Univeristy for the construction of databases for receptors and neurons to facilitate the integration of these multidisciplinary data into computational models of neurons and neuronal currents.

In summary, the information flow arrives at the axon terminal as an action potential. This action potential triggers the release of a neurotransmitter from a vesicle, whereupon the neurotransmitter binds to a receptor on ion channels, and ions cross the membrane through open channels. This influx of ions produces a synaptic potential in the postsynaptic neuron. When the integrated or total sum of the synaptic potentials exceeds its threshold, the postsynaptic neuron will fire an action potential, i.e. the neuron responds or

Fig. 2.9. John Carew Eccles. (Picture from http://nobelprize.org/nobel_prizes/medicine/laureates/1963/eccles-bio.html)

conveys information to its connecting neurons and the process continues. This process and the summing of the synaptic potentials are also known as the synaptic integration behavior. This synaptic integration behavior was first discovered by John Carew Eccles (Fig. 2.9). John was awarded a Nobel Prize for Physiology/Medicine in 1963.

How is information represented? Information is represented by patterns of activities encoded in groups of neurons (or over populations of neurons). This is also known as population encoding. Understanding the encoding of information in neural population activity is then important both for grasping the fundamental computations underlying brain function, and for interpreting signals that may be useful for the control of prosthetic devices (Singer, 2003). The computational issues include:

- What information is encoded in a population?
- How does the brain compute using populations?
- When is a population optimal?

In population encoding, each neuron has a distribution of responses over some set of inputs, and the responses of many neurons may be combined to determine some value about the inputs. In one classic example involving the primary motor cortex, Georgopoulos *et al.* trained monkeys to move a joystick in one of multiple directions towards a lit LED. Neurons in the primary motor cortex responded maximally during movements to their preferred direction, i.e. the direction neurons are trained to respond towards the lit LED, and the neuron response decreased if the animal made movements towards directions increasingly different from the preferred direction (Dayan and Abbott, 2001).

Neuroscientists found that, indeed, some neurons provide better information than others, and a population of such expert neurons has an improved signal to noise ratio or information quality. This means populations of neurons do better than single neurons. How the specific populations of neurons are assembled is thought to provide the functional elements of brain activity that execute the basic operations of information processing (Fingerlkurts *et al.*, 2004; Fingerlkurts *et al.*, 2005).

The number of neurons in a population is claimed not to be big. Some neuroscientists believe it to be not more than 100 neurons. Others believe it can be greater than 100 neurons. But the key issues is that chunks of information are encoded and stored in small populations of neurons. Here, a small population of neurons, i.e. hundreds of neurons, is considered small relative to the billions of neurons in the brain.

However, those who believe the "Grandmother Cell Theory" would say one neuron is sufficient to represent an object at the top of the highest hierarchy chain of neuron encoding. The "Grandmother Cell theory", coined by neurobiologist Jerome Levtite in the 1950s, hypothesizes that a single neuron is sufficient to store an object or a concept (including one grandmother's face). The latest support on this idea come from the study by a group of scientists from UCLA, CalTech, and MIT; see article (Quiroga *et al.*, 2005). Their results suggest an invariant, sparse, and explicit code, which might be important in the transformation of complex visual percepts into long-term and more abstract memories.

The opposing theory to the "Grandmother Cell Theory" is the "Distributed Representation Theory". The "Distributed

Representation Theory" is closer to David Hubel and Torsten Wiesel's hierarchical concept. In the "Distributed Representation Theory", it is thought a set of neurons is involved in representing a concept or symbol, and in addressing the "binding problem". The "Binding problem" refers to how different features from a picture or concept are put together.

Rick Born, Harvard Medical School, commented in his paper that the early Stanford University experiments showing that monkeys could perform the task in question (an example of a task is to push a red button when the monkey sees a red object appear) with a relatively small number of neurons, as compared with the hundred or thousand neurons one might suspect (Born and Bradley, 2005). Note: Born also concluded that MT (V5 of visual cortex) is an ideal place to evaluate models of population decoding. (I had an opportunity to visit Harvard Medical School, Rick Born's Lab, on 17 Jan 07. He highlighted that good and accurate brain experimental data took months and sometimes years to collect and validate. For example, training a monkey to do specific tasks would take about 6–9 months, and after implanting the probe, the monitoring and collection process will depend on the hours the monkey could work.)

How does the neuron's receptive field affect our visual scene? The study on the visual system's receptive field size indicated that many neurons are needed at a lower visual cortex and fewer neurons are required at a higher visual cortex to recognize a scene.

The term receptive field was coined by Sherrington in relation to reflex actions and was reintroduced by H. Keffer Hartline[b] in 1938. The receptive field of a neuron, in the visual system, can be defined as the area on the retina from which discharges of that neuron can be influenced by light. For example, a record of the activity of one particular fiber in the optic nerve of a cat shows that that fiber increases or decreases its rate of firing only when a defined area of the retina is illuminated. This area is its receptive field. By definition,

[b] H. Keffer Hartline received the Nobel Prize in 1967 for his work on the physiology and chemistry of vision.

illumination outside the field produces no effect at all. The area itself can be subdivided into distinct regions, some of which act to produce firing and others to suppress impulse in the cell.

The receptive field of the lower visual cortex is small, i.e. many neurons are needed to see a visual scene. But at the higher visual cortex, the receptive field gets larger, i.e. bigger field and more selective tuning. In V1 and V2, it is just directional, i.e. line and orientation. As it goes further up the chain the neurons become more complex and the tuning more selective. When you see an image or a scene, millions of neurons are activated in the retina. But as the image or scene is projected up the cortex layers, it is said that fewer neurons are activated due to the large receptive field.

Figure 2.10 shows how a line is seen from the lower visual area (retina) to the primary visual cortex. At the retina, many photoreceptors, each with a small receptive field, are arranged in a line. But at a higher visual area, such as the primary visual cortex (V1),

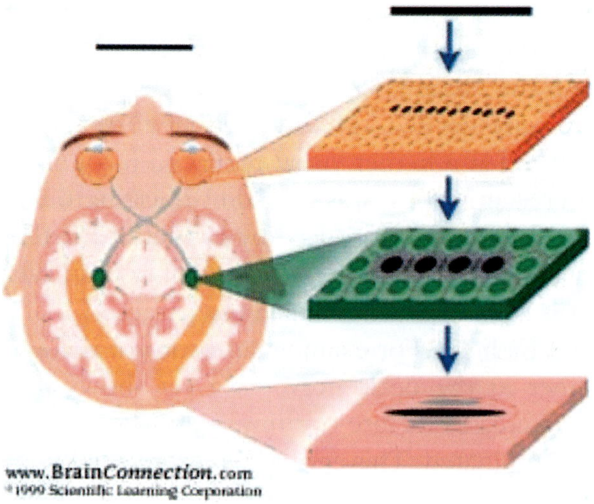

Fig. 2.10. Illustration of receptive field size from a visual perspective. (Picture from http://www.brainconnection.com)

a neuron can combine the input from the lower visual area to a line selective and orientation-specific receptive field.

How Does Synaptic Plasticity Give Rise to Learning and Memory?

Neuroscientists' believe that learning and memory are the result of long-term changes in synaptic strength, via a mechanism known as synaptic plasticity. Synaptic strength is the gain of the postsynaptic cell's response to a presynaptic input (or potential in the trans-membrane).

What is synaptic plasticity? Synaptic plasticity is the ability of the connection between two neurons, i.e. synapse to change in strength or its potential. Changes in synaptic strength can be short-term and without permanent structural changes in the neurons themselves, lasting seconds to minutes — or long-term, known as long-term potentiation, (LTP), in which repeated or continuous synaptic activation can result in a second messenger, i.e. another release of neurotransmitter, which initiates protein synthesis in the neuron's nucleus, resulting in alteration of the structure of the synapse itself. The opposing process of LTP is the long-term depression (LTD).

LTP was discovered by Professor Terje Lomo from the University of Oslo (Norway), in 1966. LTP is also regarded as the cellular basis for memory.

Hence, long term memory (LTM) is addressed (@) by long-term potentiation (LTP) under the "organization" (ORG) of synaptic plasticity (SY), i.e. LTM@LTP.ORG.SY.[c] LTP and LTD are believed to contribute to synaptic plasticity in our brain, providing the foundation for a highly adaptable nervous system. Synaptic plasticity leads to the cellular basis of memory and learning.

[c] Putting in this expression is just for easy remembering and it is not an email address. Long-term memory (LTM) is **addressed** (@) by long-term potentiation (LTP) under the **"organization"** (ORG) of synaptic plasticity (SY).

Further reading for synaptic plasticity can be found in the notes for Chapter 2.

There are several mechanisms that cooperate to achieve synaptic plasticity. These include (Gaiarsa *et al.*, 2002; Tashiro *et al.*, 2000):

- Changes in the amount of neurotransmitter released into a synapse, and
- Changes in how effectively cells respond to those neurotransmitters.
- Changes in spine morphology, i.e. production of new spines or growth of spines in a manner that produces more synaptic contact.

The synaptic plasticity due to LTP and LTD is affected by action potential. The firing of action potentials is known as spikes. And spike response with respect to stimulus has a normalization effect. The next subsection will further discuss what a spike and normalization are.

Spiking and Normalization. The timing between a presynaptic neuron and postsynaptic neuron in changing the synaptic strength is observed as a spike. The timing sensitivity of a spike's duration is on the order of millisecond.

There are various spiking patterns (or spike trains) and different single-spike shapes (for example thin-spike). Note that the speed of a spike is typically in the duration of millisecond. Figure 2.11 shows some examples of the spike patterns, namely, regular spiking (constant spike), bursting (neurons that fire in burst), and fast spiking patterns (fast firing rate). Different spike patterns are observed in

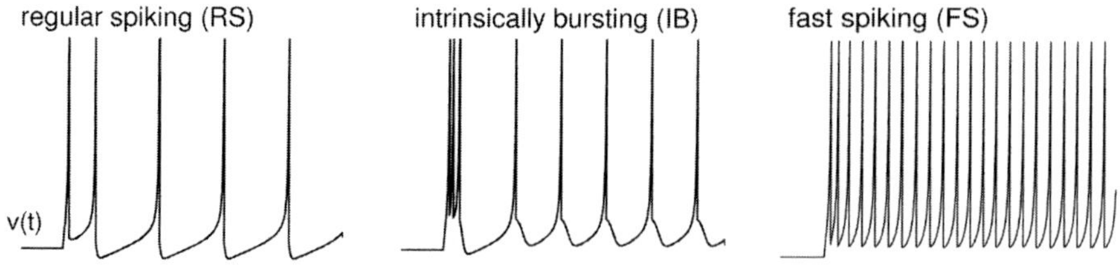

Fig. 2.11. Spike patterns.

different areas of interneuron connection. The spiking pattern is important and critical with regards to how information is stored and how information in different regions is interconnected. Spiking pattern is currently a research area and cluttered by many different hypotheses. Until better instruments and techniques are available to better prove these hypotheses, how spike patterns influence information storage will remain unknown.

It is known that the number of spikes is fewer and with slower spike rates for weak stimuli than strong stimuli, i.e. a response to strong stimulus produce a higher spike rate and a greater number of spikes. Hence, there is a need for the balance of spike responses with respect to the stimulus intensity, i.e. some kind of normalization is needed. However, it is not exactly clear how the normalization takes place in the cortex. It is certain that normalization happens at a much slower time scale than spiking, i.e. in seconds rather than milliseconds. One of the leading mechanisms for normalization is said to be the shunting inhibition in the cortex. Research work supporting shunting inhibition that led to normalization in the cortex includes (Albrecht and Geisler, 1991; Heeger, 1992; Borg-Graham *et al.*, 1998) and (Hirsch *et al.*, 1998). Shunting inhibition includes synaptic depression at active synapses and intrinsic modulation (normalization) of synaptic currents. (Carandini and Heeger, 1994) proposed that the normalization signal would take the form of a shunting inhibition driven by the pooled responses of surrounding neurons of many different preferred orientations and spatial frequencies. And this shunting inhibition, which would thus be orientation independent but increases with stimulus contrast, would increase the conductance of a cell. The conductance has a normalizing effect on the excitatory current of the cell.

However, one cannot dispute that something other than shunting may take place that somehow balances hyperpolarizing (nonshunting) inhibition with the excitatory current of the cell.

How is model normalization achieved in the cortex? One of the approaches is to use the shunting network model. This shunting network model is proposed by Professor Stephen Grossberg from Boston University and provides a simple and intuitive way of achieving normalization in the cortex. More discussion of this will be presented in Chapter 7 of this book.

Modeling the Neurons and Synapses

The idea of modeling many neurons in massive parallel connection was introduced in the 1950s and 1960s under the general field known as artificial neural networks (ANN). The subsection below provides a brief history of ANN and how the field has advanced to its current stage, and highlights the current difficulty encountered in modeling at the cellular level.

A brief historical background on artificial neural networks (ANN). In the late 1950s, Frank Rosenblatt began to explore the functional properties of small networks of neurons, which he called Perceptrons (Rosenblatt, 1962). Perceptrons works well in many experiments. But in the 1960s, many mathematicians were skeptical of Rosenblatt's conclusions because they were mainly empirically drawn. Marvin Minsky and Seymour Papert, in 1969, published the book on Perceptrons that widely determined what Perceptrons can and cannot do in the mathematical sense. Minsky and Papert's book make it clear that Perceptron architectures were much more limited in what they could accomplish than what Rosenblatt would claim. As a result, work on Perceptrons and other related neural network modeling was silent for almost a decade.

In the late 1980s to 1990s, artificial neural networks were hot topics of research again (more reading can be found in (Ng, 1997) where an extensive survey was conducted in 1994). The field was re-ignited by various factors, some of which include: publication of the books "Parallel Distribution Processing" by Rumelhurt and McClelland 1986/87 (Rumelhurt and McClelland, 1986a,b) (Rumelhurt and McClelland, 1987a,b), interest from DARPA[d] to re-examine neural information processing capability (DARPA neural

[d] DARPA stands for Defense Advanced Research Projects Agency. DARPA is the central research and development organization for the USA Department of Defence.

network study), and disappointment faced in traditional AI techniques for performing higher cognitive tasks. In late 1990s, the field slowed down. This was because most models of the neural network have shown static learning and could not perform beyond the promise as claimed. However, research persists with more accurate models of the neural networks. These include modeling the detailed synaptic dynamics, neuron interconnection, the pathway, cortex architecture, neural cortical column structure (more information of neural cortical column in Chapter 7), excitatory and inhibitory effect, horizontal interaction, etc. However, difficulties still persist and many adopt a simplified approach to a very precise and accurate model. To illustrate this difficulty, we will consider modeling at a cellular level, i.e. modeling of the neurons and synapses.

Modeling at a cellular level. Modeling the neurons and synapses with precise accuracy can be very complicated. One example is the Hodgkin–Huxley cell membrane model. The Hodgkin–Huxley model, proposed by Alan Lloyd Hodgkin and Andrew Huxley in 1952, represents the biophysical characteristics of the cell membranes and describes how action potentials in neurons are initiated and propagated. Despite being a biophysically accurate model, it is computationally prohibitive for simulating many neurons in real time, and many parameters need to be precisely determined from the real biological neurons.

Other computationally effective models based on neuron and synapses behaviors are available, such as the integrate-and-fire model and spiking model. However, these are typically simplified models of neurons. Many models were proposed to produce so-called realistic spiking and bursting dynamics exhibited by cortical neurons. One of these models is the simple spiking model proposal by (Izhikevich, 2003). Izhikevich claimed that the simple spiking model provides rich spiking and bursting dynamics. The details of the Hogkin-Huxley equation and the simple spiking model are presented in the notes on Chapter 2.

Summary

Neurons communicate with one another through axons and dendrites via the synapse. At the synapses, information is passed from one neuron's axon to the next neuron's dendrites. Throughout our entire life, the brain responds to experiences by adjusting the synaptic strength and by changing the physical pattern of synaptic connections between neurons. Thus, information can be stored by the nervous system in the form of altered structures of the synapses, and/or by the formation of new synapses and the elimination of old ones. A permanent change in synaptic strength, which leads to long-term retention of information in memory, is due to the long-term potentiation (LTP). The opposing process is long-term depression (LTD). Both the LTP and LTD influence how we learn and store information.

The firing of action potentials is known as spikes. The number of spikes and the spike rate determine the intensity of the stimulus. It is generally believe that normalization takes place in the cortex to avoid neuron saturation.

The number of neurons involved in storing specific information is probably small[c] at the cortex. The debate on whether a single neuron or a large set of neurons for specific information storage will become clear only when better measuring instruments can be used to perform the experiment.

Modeling of neurons using a biologically accurate model, such as the Hodgkin–Huxley cell membrane model, is computationally complex. Other computationally effective models based on neuron and synapses behaviors, such as the integrate-and-fire model and spiking model are commonly used. However, these models typically simplify the modeling of neurons and may not lead to an accurate model of the true neurons.

---

[c] I regard less than a thousand neurons as small relative to the billions of neurons in the brain.

Chapter **3**

THE CORTEX ARCHITECTURE: THE BUILDING BLOCK OF INTELLIGENCE

"*Innate mechanisms endow the visual system with highly specific connections but visual experience early in life is necessary for their maintenance and full development. Deprivation experiments demonstrate that neural connections can be modulated by environmental influences during a critical period of postnatal development. We have studied this process in detail in one set of functional properties of the nervous system, but it may well be that other aspects of brain function, such as language, complex perceptual tasks, learning, memory and personality, have different programs of development. Such sensitivity of nervous system, to the effects of experience may represent the fundamental mechanism by which organism adapts to its environment during the period of growth and development.*" Torsten N. Wiesel, Dec 1981 Nobel lecture.

Scientists around the world have discovered that genes influence the size and overall structure of the cortex, including the regionalization and specialization of the brain area. Activity stimulus and experience through sensory input further build the cortex to specific functions or sharpen the functions. Some of the specialized functional regions in the cortex are discussed in Chapter 1.

The cortex surface is highly folded, i.e. extensive wrinkling and convolutions. The cortical folding appears to have advantages, such as increased surface area and volume ratio, and provide for short connections between different functional regions in the cortex. This interconnection is known as cortico-cortical connections.

Hence, the brain can have a large number of neurons per unit volume and increase in capacity and efficiency with a relatively small head size compared to the body size. Understanding the functional significance of these folds is one of the big open questions in neuroscience.[a]

From the computational modeling perspective, it is interesting to know how the cortex is structurally arranged and what are the underlying functional layers that enable us to recognize, discriminate, understand, and perform decision-making in a relatively short time span. This chapter will discuss the following issues:

- How is the cortex structurally arranged?
- What are some of the factors that influence the cortex architecture?
- Discuss the computational model of the cortex architecture.
- Present a case study on why the brain can discriminate and recognize pictures at a glance.
- How the prefrontal cortex performs decision-making.

How is the Cortex Structurally Arranged?

Early experiments on the visual processing in the cortex has shown the hierarchical organization of the visual cortex in the brain. Hubel and Wiesel (Hubel and Wiesel, 1962) were probably the first to demonstrate that the visual system in the brain is designed in hierarchical structure and with highly specific connections. They also proposed the notion of the hierarchy feature extractors model from simple to complex cells for the visual cortex. Ungerleider and

[a] For example, how the folds develop and decay in the cortex, and what is the significance of the course fold and the fine-grained folds.

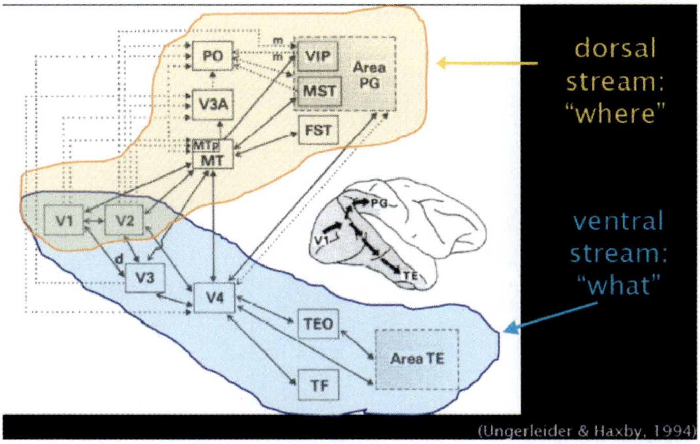

Fig. 3.1. Picture showing the two distinct visual pathways — the ventral stream and the dorsal stream (Ungerleider and Haxby, 1994).

Haxby in 1994 further illustrated that the hierarchical organization of the visual cortical area can be further broken down into two distinct hierarchically organized pathways, namely the ventral and dorsal streams (Ungerleider and Haxby, 1994).

The two distinct pathways (Fig. 3.2), namely, the dorsel stream, is shown in green, and the ventral stream, is shown in red (Ungerleider and Haxby, 1994). Leslie G. Ungerleider, from the National Institute of Mental Hospital, is BU/CNS Celest Distinguished guest speaker in December 2006. Their figure has been heavily cited and studied. Most visual cortex models are developed based on this diagram.

The ventral stream of visual cortex is often referred to as the "what pathway" because it is widely believed to be associated with form recognition and object representation (Kandel *et al.*, 2000). The dorsal stream of the visual cortex is often referred to as the "where pathway" as it is often believed to associate with the direction and orientation of an object location.

Figure 3.3 shows another perspective of the hierarchal organization of the visual cortex in simple block form. Figure 3.3 on the right illustrates the fact that the visual cortex is physically arranged as patches on a planar cortical sheet (Dean, 2005).

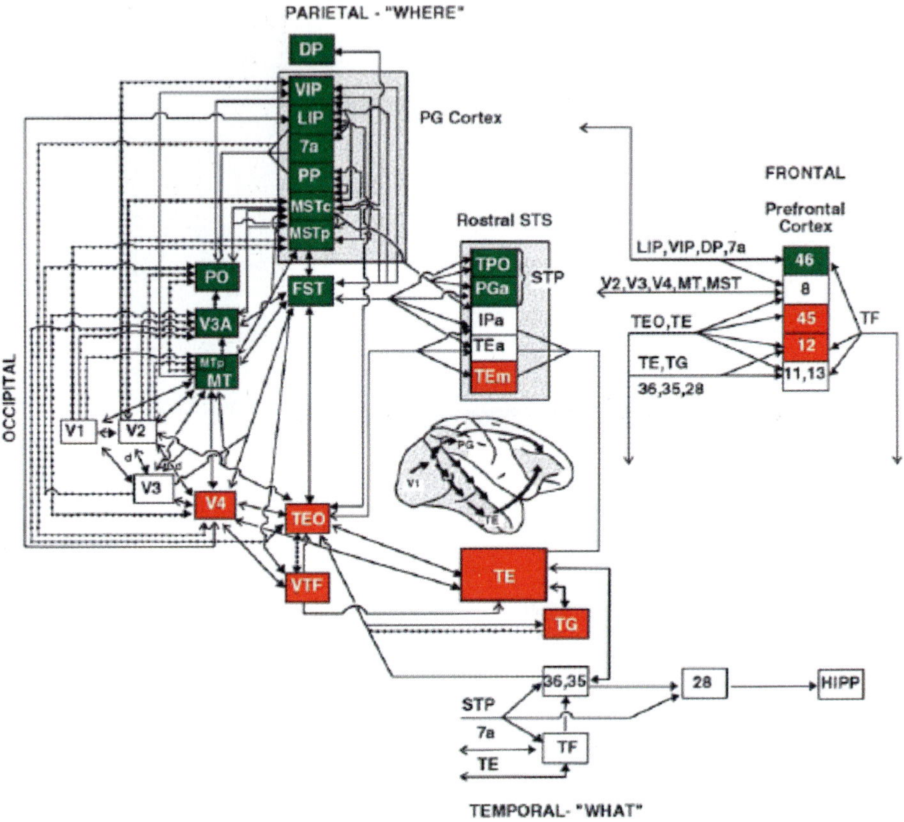

Fig. 3.2. Picture showing the detailed breakdown of the visual processing pathway, feedforward and feedback path are show by the arrows.

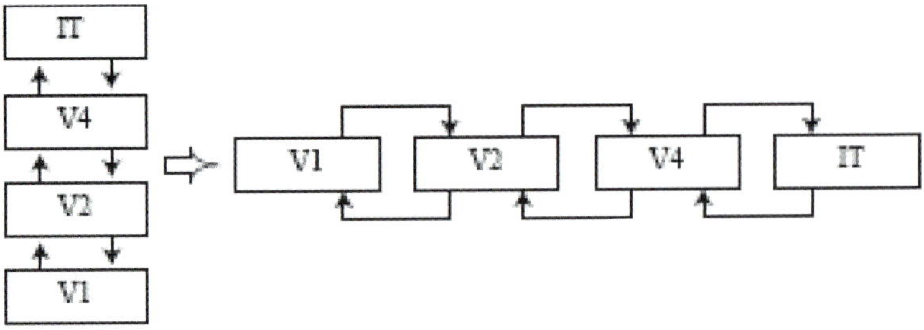

Fig. 3.3. Hierarchical organization of the visual cortex.

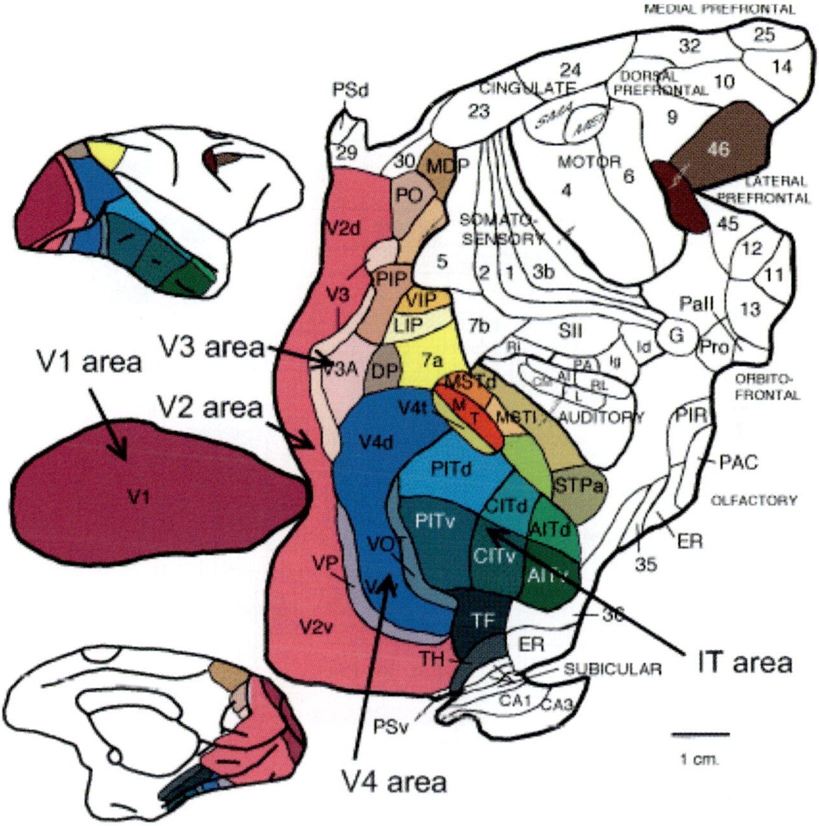

Fig. 3.4. A flat brain of the Macaque monkey. (Picture adapted from (Felleman and Van Essen, 1991))

Figure 3.4 shows the flat brain of the Macaque monkey. The general four areas for the visual cortex, namely, the V1, V2, V3, V4, and IT areas are shown in bold letters. The detailed interconnection of the visual cortex, from LGN to the prefrontal cortex, is shown in Fig. 3.5.

It is clear that the primary visual cortex is divided into hierarchical regions (Figs. 3.2, 3.3, and 3.5). Regions that are higher in the stack capture general visual features, while those lower in the stack capture specific features. Other regions in the cortex

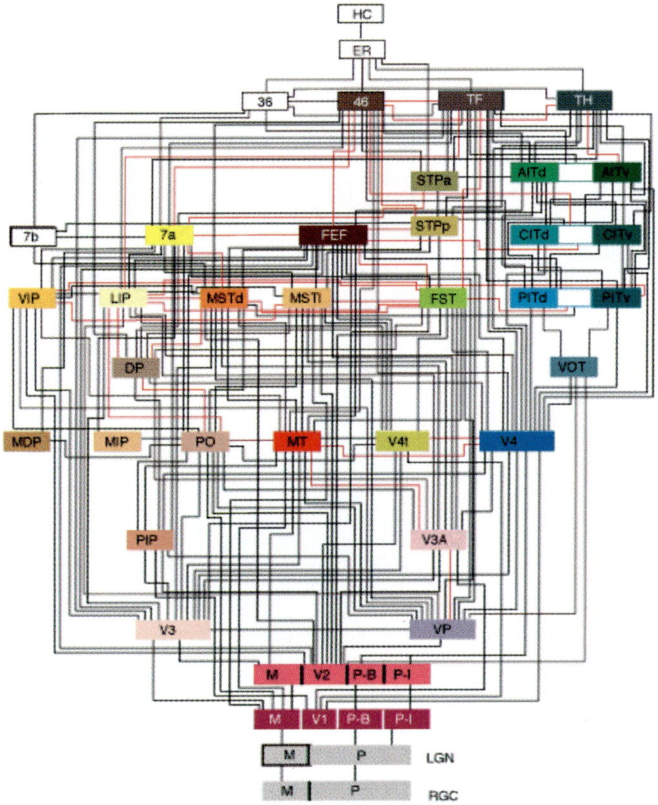

Fig. 3.5. Schematic diagram of the visual cortex arranged in hierarchical order from the analysis of the macaque monkey's brain. (Picture from Felleman and Van Essen, 1991.)

are also shown to be arranged in hierarchical forms. For example, the saccade mechanisms that lead to the cerebral cortex and the basal ganglia regions are also shown to have hierarchical connections.

Figure 3.6 shows the hierarchical organization of possible saccade mechanisms. Effective connections to the SC (superior colliculus) from the cerebral cortical areas and the basal ganglia (i.e. SNr) enable the selective control of saccades especially by suppressing unwanted saccades. (Hikosaka *et al.*, 2000)

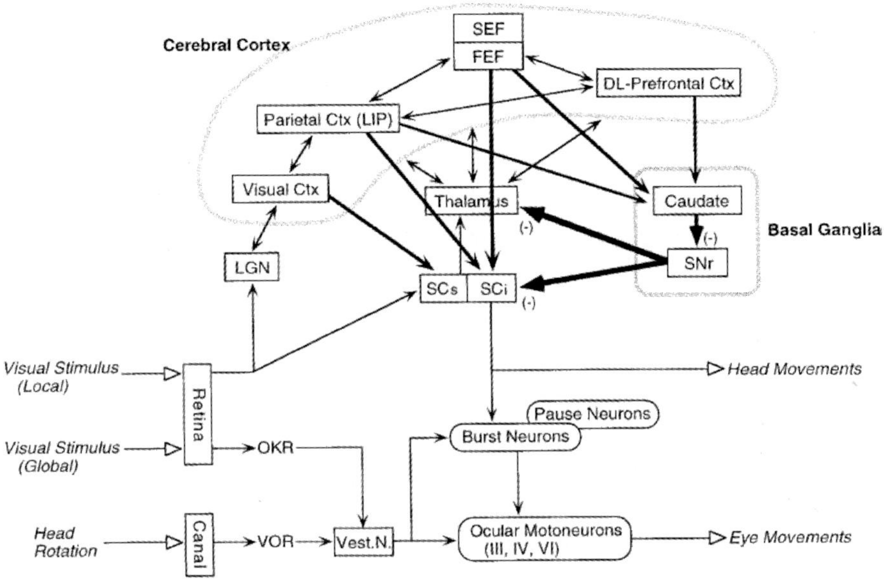

Fig. 3.6. Hierarchical organization of saccade mechanisms.

Other Design Factors in the Cortex

Besides hierarchical structure, the following are some other design factors that influence the cortex architecture.

How do different regions in the cortex interact? Recent studies have also shown that the cortex relies on specific input pathways, processing networks, and output projections. "*Pathways and networks confer representation*" (Sur and Leamey, 2001).

> "*The early development of pathways to and from a cortical area involves the expression of specific genes which regulate molecules that guide axon projections. We have used high-density DNA microarrays to identify genes that are differentially expressed between sensory areas of cortex in neonatal mice.*" (Leamey *et al.*, 2002).

This suggests specific hierarchical pathways in the cortex. The cortex also continually changes its fine functional role according to the influence of attention, experience, activity, expectation and perceptual tasks. This further suggests that the properties of any cortical area are dynamic, experience-dependent, and subject to top-down influences and horizontal interaction (Gilbert *et al.*, 2000; Grossberg 2003a; Li *et al.*, 2004).

> "*The cerebral cortex is subdivided into discrete functional areas that are defined by specific properties, including the presence of different cell types, molecular expression patterns, micro circuitry and long-range connectivity. These properties enable different areas of cortex to carry out distinct functions. Emerging data argue that the particular structure and identity of cortical areas derives not only from specific inputs but also from unique processing networks.*" (Majewska and Sur, 2006)

> "... *Visual Cortex, like many parts of perceptual and cognitive neocortex, is organized into six main layers of cells, as well as characteristic sub-lamina. ....these layered circuits help to realize processes of development, learning, perceptual grouping, attention and 3D vision through a combination of bottom-up, horizontal and top-down interactions...*" *Stephen Grossberg, Invited article for Behavioral and Cognitive review* (Grossberg, 2003a).

Hence, the basic cortex architecture consists of multiple feedforward and feedback hierarchically layered networks. There are also pathways that direct specific connections with other cortices. Some cortical regions have unique properties depending on the type of sensory input. The cortex connection is further built up with influence from top-down, bottom-up, and horizontal interaction. The layer interconnection can be at distance location.[b] Hence, the top-down influence can be from a different region depending on the context, and bottom-up information may drive a distant layer in the cortex. The cortex structure continues to build up from this

---

[b] Distance location refers to a long axon connecting to another cortical area in the brain.

basic "scaffold" architecture depending on external activity and its specific properties to form specific cortex functions.

Is the cortex a primary memory system? "The brain can keep the eye still because it stores a memory of the eye position." H. S. Seung (MIT/BCS) (Seung, 1996).

Since the 1990s, most neuroscientists support or agree, as I understand from conversations with neuroscientists and from reading published papers, that the brain's cortex mainly involves memory systems.[c] If this is so, how are these memories interconnected such as to enable us to understand and recognize object and concept in invariant time and space? What is the memory temporal-spatial representation of information? How is the information recalled and retrieved at the "appropriate" cue? Cues are based on many factors such as timing, events, and related pattern. In computational terms, this could be related to a "pointer".[d] How can we have a "dynamic pointer"? In some sense, we do not know how the "dynamic pointer" in the brain is designed and invoked, and sometimes its directing is not constant or cannot be defined in a consistent way. More importantly, sometimes, we wonder why these "dynamic pointers" are sharp, directed, and pointed.

Brain performs a fusion process. In (Ng, 2003), I discussed that fusion is an integration of multiple sources of information in the form of competitive and/or complementary ways. Studies till now have supported the existence of both the competitive and complementary interaction of the information process in the brain. Hence, in some sense there is "fusion" performed in the brain.

For example,

- Ventral ("what") and Dorsal ("where") stream working together in a complementary manner help us to view, classify,

[c] Some scientists may not fully agree that the brain is mainly composed of memory systems.

[d] In software coding, a pointer is a special kind of variable that holds the address of another variable.

and recognize the orientation of an object in a natural scene. In each stream, the six-layer cortical structure interconnections (the excitatory/inhibitory and feedforward/feedback loop) are performing competitive actions to achieve its intrinsic goal.

- The prefrontal cortex that performs the function of working memory is able to piece together information from various parts of the cortex in the brain to process information, such as forming a good sentence in our speech.

Detailed studies of the visual cortex at the neuron level also confirm that different specialized neurons working in complementary and competitive manners enable us to see the image laid in front of our eyes.

Computational Model of the Cortex Architecture — Model After the Hierarchical Structure

This section presents some of the computational models of cortex architecture. Most current work has been done on the visual cortex. Researchers in the 1990s have proposed a hierarchical structure for modeling the brain's cortex, and have applied machine learning techniques to update the connecting weights in the hierarchical model. More lately, researchers have re-proposed the same idea and built models on hierarchical manner. For example, Riesenhuber and Poggio proposed the Hierarchical Feedforward architecture (Riesenhuber and Poggio, 1999), Lee and Mumfold suggested the Hierarchical Bayesian Model with Particle Filter (Lee and Mumfold, 2003), and Jeff Hawkin talked about Hierarchical Temporal Memory (Jeff, 2004). This section will discuss some of these computational models of the cortex in hierarchical structure.

> "*The recognition of visual objects is a fundamental cognitive task performed effortlessly by the brain countless times every day while satisfying two essential requirements: invariance and specificity. In face*

recognition, for example, we can recognize a specific face among many, while being rather tolerant to changes in viewpoint, scale, illumination and expression." (Lee and Mumfold, 2003)

Professor Tomaso Poggio, Director of the Center for Biological and Computational Learning at MIT Brain and Cognitive Science (BCS) Department has a group of researchers who demonstrated an ultra-fast object recognition system using a hierarchical feedforward architecture in 2006. They suggested that hierarchical feedforward architecture (without feedback) may explain the effect of human "immediate recognition" — the initial phase of recognition before eye movements and high-level processes can play a role.

Figure 3.7 shows the model of the hierarchical feedforward visual cortex. Serre's research work in MIT modeled the ventral stream of the visual cortex. Figure 3.7 on the right shows the modeling of simple and complex cells (Serre *et al.*, 2006, 2007a,b). Serre

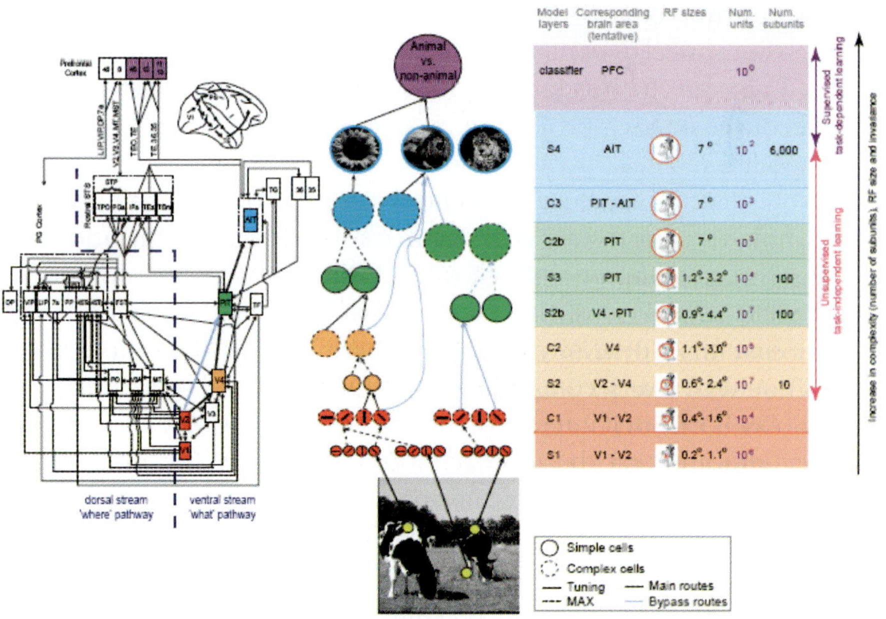

Fig. 3.7. Model of the hierarchical feedforward visual cortex from Serre's research. (Picture from Serre *et al.*, 2006.)

modeled the details based on the different functional layers described in the ventral stream. Supervised training was performed at the prefrontal cortex layer and unsupervised learning was performed at lower level such as V1 to IT layers.

Serre *et al.*, in their paper, claimed that their hierarchical feedforward model behaves similarly with the human observer across all four animal categories in the ultra-rapid categorization of 50 ms in recognizing animals in the natural scene. (Animals in the natural scenes constitute a challenging class of stimuli due to large variations in shape, pose, size, texture, and position in the scene.) The MIT team further tested and showed that the model is robust on rotated images not re-trained before.

However, they reported that at 80 ms, the human outperformed the model. They concluded that it is an open question whether the somewhat better performance of humans for longer than 80 ms is due to feedback effects mediated by back-projections, i.e. top-down influence.

Professor Tomaso Poggio, in 2007 Scene Understanding Symposium, declared that the feedforward ventral stream model was completed; he was looking for the next stage of incorporating the feedback or back-projection process and aimed for image inference abilities.

Note that the feedforward ventral stream is only a small part of the whole visual pathway. With feedback influence, consideration of object recognition in depth, classifying different objects in the natural scene, stability of the network with positive feedback, and addressing the dorsal stream (the "where" issue), the whole network could be a very complicated one. Other potential gaps include the global horizontal interaction not included in the model. Current debate indicates that the approach looks like a convolution method and needs to resolve the shift invariance part. The simple cell layer uses the Gabor filter. The Gabor function has long been associated with visual processing. (Stork and Wilson, 1990) pointed out that the Gabor function cannot fit to single-cell receptive fields and they concluded that there are insufficient theoretical demonstrations and experimental data to favor the Gabor function over any number of other plausible receptive-field functions. Despite this, the Gabor function remains popular and is used in the Serre *et al.* model. The model is only tested on simplified categorization tasks such as animal-versus non-animal or face versus non-face. It has not been used to recognize particular faces or specific animals' identification.

Professor Stephen Grossberg's research team from Cognitive and Neural System (CNS) Department in Boston University has studied and worked on cortical network architecture and the resonance theory for many years. Their team has considered both the ventral stream ("what") and the dorsal stream ("where") in their modeling of the visual pathway. His group is progressing up the chain from LGN–V1–V2 to MT/MST and AIT with feedforward, feedback, inhibitory signal, and horizontal interaction, in one holistic consideration. Their model, in some sense, has a closer resemblance to the biological theory and is careful in considering each connection to address the different features involved in capturing the vision scene.

Figures 3.8 and 3.9 show the feedback, excitatory, and inhibitory signal consideration in the design of the partial visual cortex. The design includes the six-layer cell interconnection.

The model (Figs. 3.10 and 3.11) explains how surface attention interacts with eye movement for object recognition from different view angles. Fazl indicates that this interaction does not require prior knowledge of object identity. The modules in the model conform to brain regions in the "what" and "where" cortical streams of the visual system. The "what" stream learns a spatially-invariant and size-invariant representation of an object, using bottom-up filtering and top-down attention mechanisms. The "where" stream computes indices of object location and guides attentive eye movements. Preprocessing occurs in the primary visual areas, notably log-polar compression of the periphery, contrast enhancement, and parallel processing of boundary and surface properties (Fazl *et al.*, 2005b). This model is known as ARTSCAN.

Fazl in CNS has completed the above model and demonstrated its invariant representation and the abilities to perform 3D object categorization. He demonstrated that the ARTSCAN model has

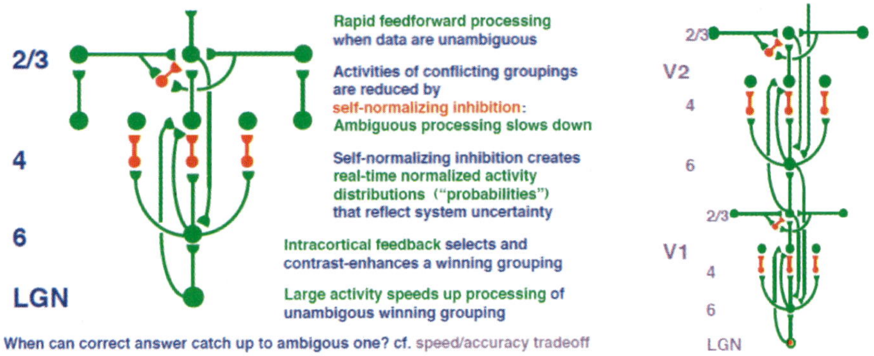

Fig. 3.8. (*Left*) shows the layer connection in the LGN and V1; (*Right*) shows the extension from LGN-V1 to V2.

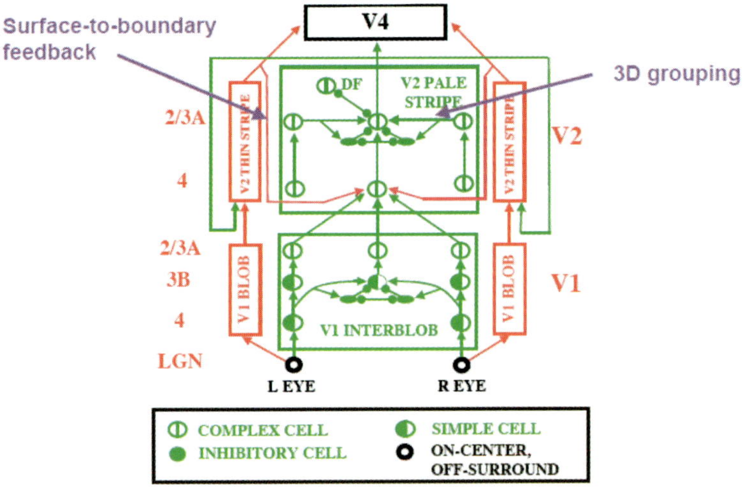

Fig. 3.9. The extension from LGN to V4.

a performance of above 95% correct invariant object recognition after real-time incremental learning controlled by attention shifts and active eye movements. Invariance here refers to the object that can still be recognized when it is presented in different orientations, depths, and sizes. Note that the object invariant view is limited within the training of object orientation and size.

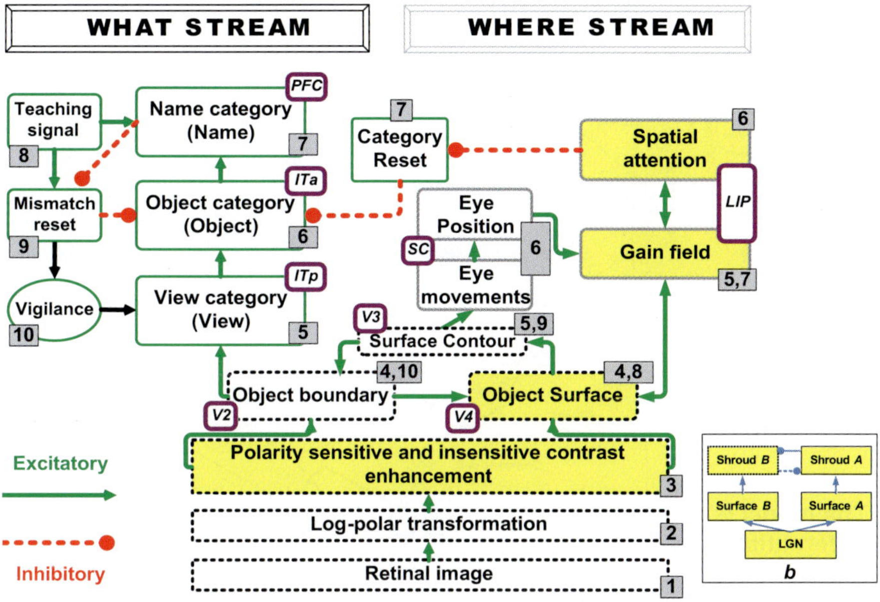

Fig. 3.10. Schematic diagram showing the connection of the dorsal stream (where) and the ventral stream (what) modeled for 3D object recognition. (Picture from Dr Fazl, BU/CNS)

Current experiment trains objects at double their size and orientation at +45 degrees to –45 degrees (Fazl *et al.*, 2005a,b, 2006).

Supervised and unsupervised learning are performed. Supervised learning is done at the higher cortex (prefrontal cortex area) and unsupervised learning is done at the low cortex areas. The reason for supervised learning in the prefrontal cortex is that the object could only be trained (or presented) at a complex level, which is at the prefrontal cortex and the intermediate layers (V2–IT) where the object is broken down into its components, and unsupervised learning is performed (clustering of the components). This is the same logic for Serre *et al.* in their work using supervised learning at the top layer and unsupervised learning at the layer below this top layer. In Serre *et al.*, the extraction at each layer takes the strongest average signal. Fazl's work demonstrated

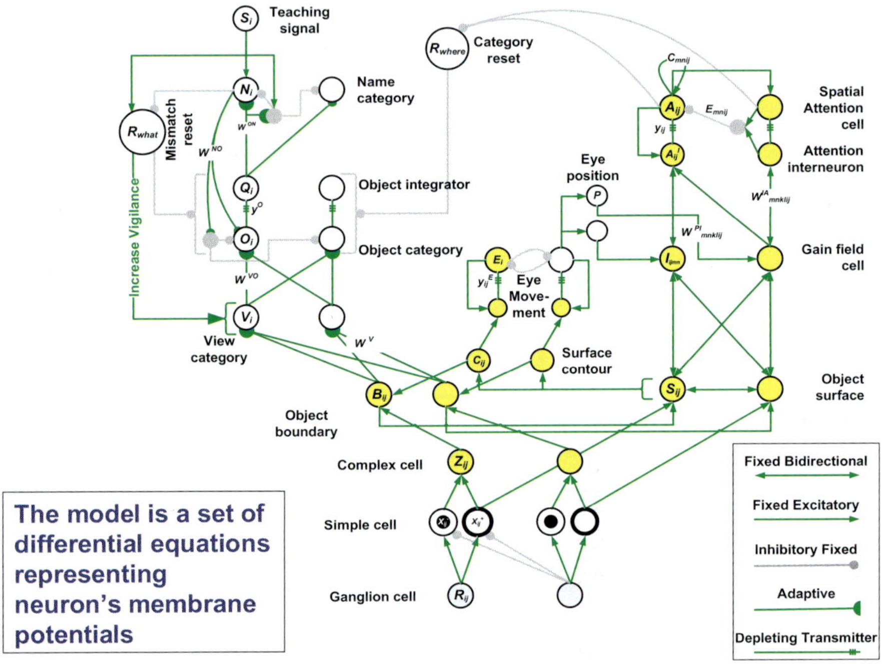

Fig. 3.11. The circuitry connection of the dorsal stream (where) and the ventral stream (what) modeled for 3D object recognition. (Picture from Dr Fazl, BU/CNS)

all the major components for signal flowing forward and with feedback, according to the cortical connection.

Figure 3.12 shows the several variations of Hierarchical Temporal Memories (HTMs) proposed by Jeff Hawkins and Dileep George from Numenta Inc. Jeff indicated that HTMs are similar to Bayesian Networks (BNs); however, HTM differ from most BNs in the way that time, hierarchy, action, and attention are used.

The HTM receives the pattern coming from the senses. It then creates a set of beliefs. Each belief is back-propagated (BP) up the HTM. The CPT (Conditional Probabilistic Table) is created on-the-fly by learning the quantization points. Note that quantization points at the bottom nodes are created by the spatial pattern presented and are propagated upward. The quantization

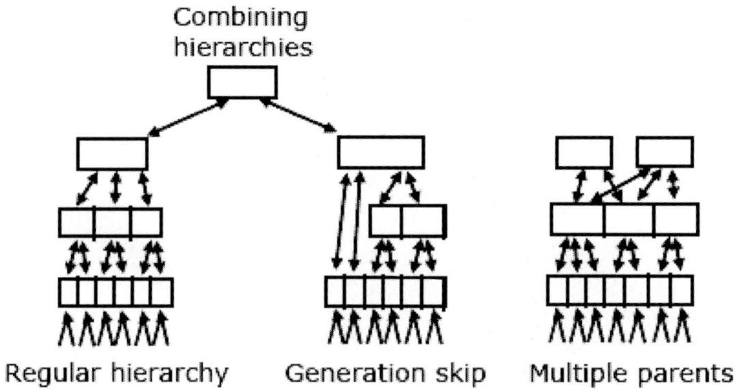

Fig. 3.12. Hierarchical Temporal Memories (HTMs).

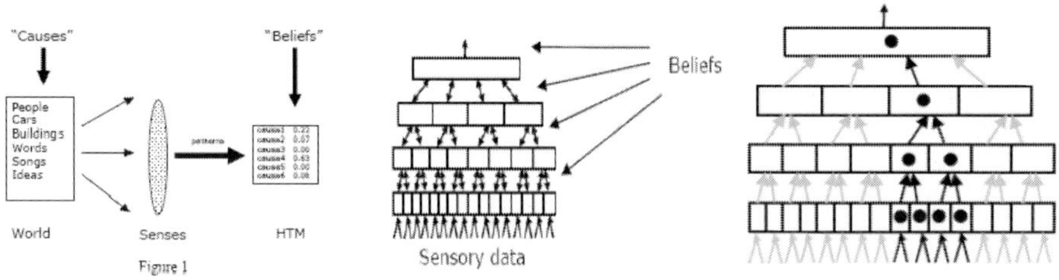

Fig. 3.13. The representation of causes and beliefs in the HTM.

function itself is the CPT in Jeff's HTM. This way of creating CPT differs from traditional BNs. Everything the HTM learns is stored in memory matrices at each node. These memory matrices represent the spatial quantization points and sequences learned by the node.

In perceptual learning, Ahassar and Hochstein proposed a reverse hierarchy theory of explicit feedback connections and implicit feedforward connections for vision (see Fig. 3.14) (Ahissar and Hochstein, 1998; 2004). This structure has the same notion as a simple to complex cell hierarchy. It indicates that higher up the hierarchy, the complex cell represents abstract information of the world and it helps in perceptual learning via feedback connections from the high-level. The early explicit perception is a post-processing

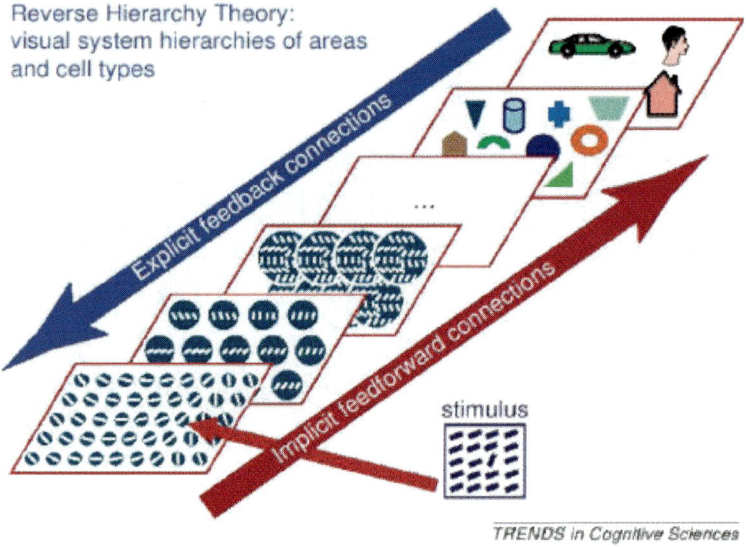

Fig. 3.14. The reverse hierarchy theory. (Picture from Ahissar and Hochstein, 2004.)

high-level view with spread attention. Later, attention is focused on specific low-area details.

For object recognition, other related work that claim to work extremely well include:

- Constellation models have been shown to learn to recognize many objects (one at a time) using an unsegmented training set from just a few examples (Weber *et al.*, 2000; Fergus *et al.*, 2003; Fei-Fei *et al.*, 2004). The model used learnt incrementally in a Bayesian manner and made use of prior information (Fei-Fei *et al.*, 2004).
- Multiplayered convolutional networks were shown to perform extremely well in the domain of digit recognition and face identification (Fukushima, 1980; LeCun *et al.*, 1998, 2004; Chopra *et al.*, 2005).
- Hierarchical Memory system with feedback. Another approach using a self-organizing map principle with focus on feedback connection (Sit and Miikkulainen, 2006).

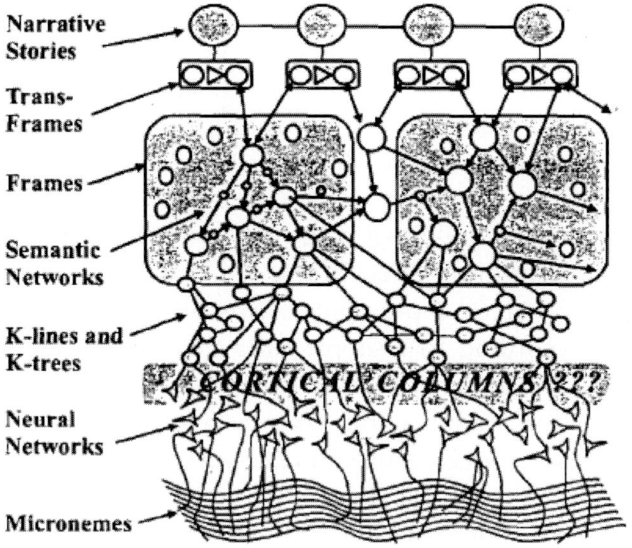

Fig. 3.15. Professor Marvin Minsky's way of representing multiple connections in the brain. (Picture from "The Emotional Machine" by Marvin Minsky.)

Figure 3.15 shows an attempt by Professor Marvin Minsky, MIT media laboratory, on how a brain might organize its multiple hierarchical connection to represent knowledge (Minsky, 2006).

> "*It also seems unlikely that our representations are arranged quite so hierarchically. In biology, new structures usually originate as duplicate copies of older ones, and this often results in orderly layers. However, because brain cells are so peculiarly able to make connections to distant places, they can more easily evolve less hierarchical organizations.*" Marvin Minsky (Minsky, 2006)

Case Study — Why the Brain Can Discriminate and Recognize Pictures in a Glance

A human can perform significant tasks in much less time than a second, such as recognizing an object in a natural scene. Professor Mary Potter, in her seminar talk (November 2006 at MIT/BCS), indicated that a picture can be understood by a human in about 100 millisecond (ms). This is supported by her experiment. Now, how far could the neuron signal travel in 100 ms? This question was raised and it triggered me to find out more. Given that the synaptic transmission delay is about 1–10 ms, a neuron can travel at the speed of 10–100 m/s.

Now, let us take the visual pathway. Figure 3.16(A) shows the hierarchical organization of cortical areas. This model presents the different cortical areas of the primate's visual system, staged at different levels according to a simple rule: areas at low-order stages send feedforward connects to the upper levels, whereas high-order areas send feedback connections to areas at a lower level (Bullier, 2001). And Fig. 3.16(B) shows the latencies of visual responses of neurons in different cortical areas.

So if a picture is presented in 100 ms, the signal of the "image" of the picture received could have traveled to the inner cortex of the brain (all the way to the prefrontal cortex). Hence, within 100 ms, we can recognize a picture. One neuron potentially could be connected to another 10 000 other neurons as it moves up different cortical areas. (Although the signal has traveled through many neurons, it must also able to retrieve the information from the LTM to process within 100 ms. Mary Potter coined the term conceptual short-term memory to explain rapid comprehension of new stimuli such as pictures or sentences that require retrieval from the LTM. She explained that there is a need to retrieve the conceptual information from the LTM for processing. (Note: A question was asked on what is the difference between working memory and conceptual short-term memory. Her answer was conceptual short-term memory is a more précise definition.)

The ability of the brain to search at specific neural pathways looks critical here. It is well-known that certain regions of the brain contain certain information. For example, for face recognition, the part of the brain that contains the facial image is in the fusiform face area (FFA). How would the brain know where to search? Or is it a pure brute force search due to its massive parallel connections?

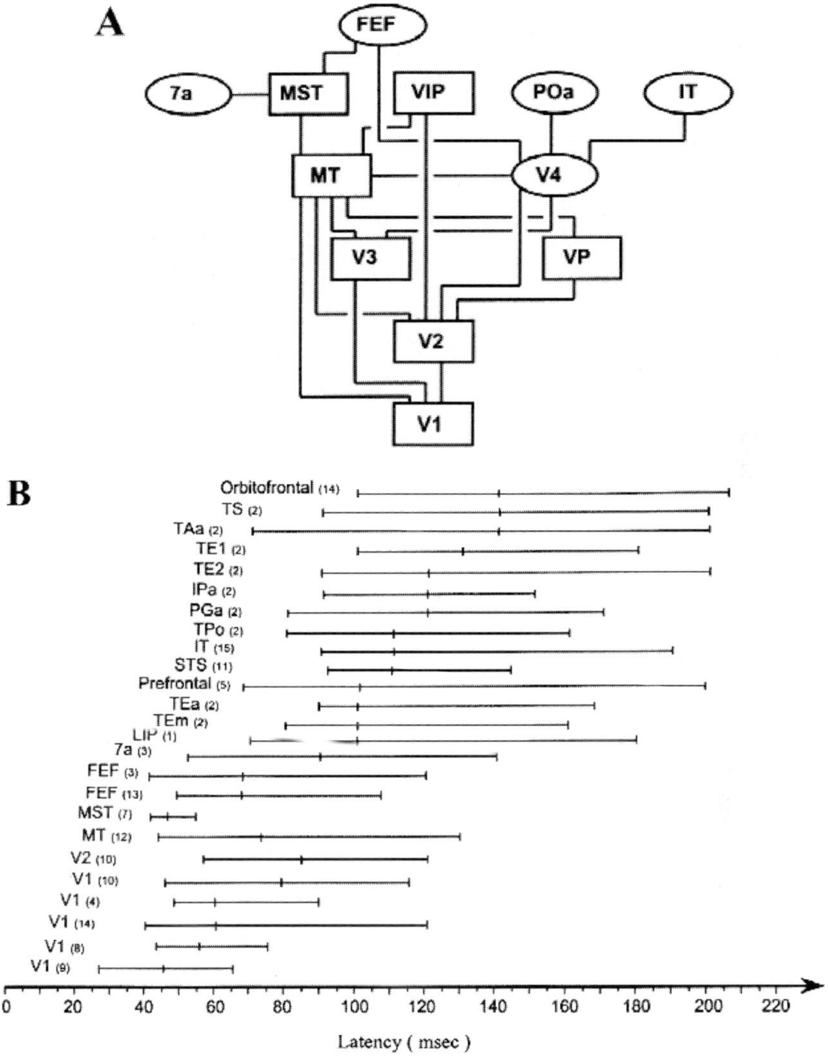

Fig. 3.16. (A) Hierarchical organization of visual cortical areas (Bullier, 2001). (B) The latencies of visual responses of neurons in different cortical areas.

For each area (Fig. 3.16(B)), the central tick marks the median latency and the extreme ticks the 10–90% centiles (Bullier, 2001). (See also Fig. 3.17, an approximate time scale from the monkey experiment.)

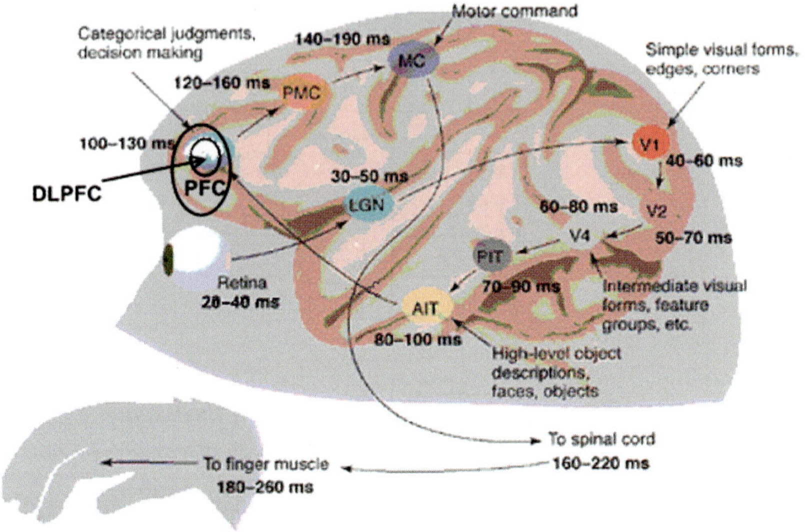

Fig. 3.17. An approximate time scale from visual to motor response from a monkey experiment. The diagram also shows an estimated location of the dorsolateral prefrontal cortex (DLPFC) in the brain. (Picture adapted from (Thorpe and Fabre, 2001))

So for about 100 ms the signal from the retina would have reached the prefrontal cortex, whereupon it would pass the AIT where high-level objects are stored. The MIT team modeled the ventral stream using a hierarchical feedforward architecture. They suggested that there is probably not much feedback for fast discrimination (Serre *et al.*, 2006).

How Does the Prefrontal Cortex Perform Decision-Making?

The prefrontal cortex has been commonly associated as the place where working memory resides and decisions are made. The motor pathways are link to the prefrontal cortex as shown in Fig. 3.17.

Figure 3.17 shows the possible path the signal will travel for a brain to make a simple decision, say finding an object in a cluttered picture, and respond with the subject pressing a button. Monkey and human are able to respond at about 200–400 ms (Thorpe and Fabre, 2001).

Ungerleider and her team use functional magnetic resonance imaging (fMRI) and a categorization task in which subjects decide whether an image presented is a face or a house to test where and how in the brain this decision-making computation might be performed. (Heekeren *et al.*, 2004, 2006). She concluded that:

- An increase of sensory network activity or attention occurs when the information presented is noisier. This implied that greater noise results in more attention focus.
- The prefrontal cortex has a general decision-making function independent of the stimuli and response modalities.
- When subjects made categorical decisions about degraded face and house stimuli, the activity within the left dorsolateral prefrontal cortex (left DLPFC) is greater during easy decisions than during difficult decisions. Ungerleider indicated that posterior DLPFC looks more like the areas in the brain for difficult decision-making process.

Ungerleider's experiment also showed that 75% of the people tested performed better at the noisy face recognition process when prior face information was given.

Stepwise activation — Hierarchal sequence in problem-solving. The brain relies on stepwise activation to make difficult decision (Dobbs, 2006; Bunge *et al.*, 2005). David Badre (formally in UC Beceley, now in MIT/BCS) performed an experiment, which showed that as test subjects faced more variables in a given problem, they recruited first the premotor cortex (first area just behind the prefrontal cortex), and then as the context became more important, the pre-premotor cortex (second area), a section of the prefrontal cortex directly in front of the premotor cortex, was used. And a third area, the lateral prefrontal cortex, would kick in if solving the puzzle required weighing past events or ongoing goals. And it has been observed that the lateral prefrontal cortex (or "advanced" areas in solving problems) does not light up unless the ones preceding them had activated. Once the lateral prefrontal cortex is "on", they stayed active longer, as if monitoring the more complex situation. The fourth area is the frontal pole, at the very front of the lateral frontal lobe. Badre used imaging studies to show that these four areas operate in hierarchal sequence as the subjects solve an increasingly difficult series of puzzles, i.e. for simple problems the premotor cortex is activated, and for the most complicated puzzles the frontal pole (frontmost part of the prefrontal cortex) consistently lit up.

Summary

From the study of the visual cortex and other cortical regions in the brain, it is clear that the cortex is divided into hierarchical regions. Each region has a layered structure arranged from a simple layer consisting of simple cells to a complex layer consisting of complex cells. Within the hierarchies, computationally, supervised learning seems to be more appropriate for the higher cortex, such as the prefrontal cortex, and unsupervised learning is done at the lower cortex such as the lower sensory cortex.

These multiple memory systems are implicitly located all over the cortex. The basic cortex architecture consists of many specific pathways that link different regions in the hierarchies together. These pathways and their flow enable influence from top-down information (feedback), bottom-up (feedforward), and horizontal interaction. The layer interconnection can be at distance location. Hence, the top-down influence can be from a different region depending on the context, and bottom-up information may drive a distant layer in the cortex. The cortex structure continues to build up from this basic "scaffold" architecture depending on external activity and its specific properties to form specific cortex functions. The cortex also performs fusion-like processes that involve complementary and competitive processes. Research as of now also supports the cortex as being primarily composed of multiple memory systems.

Experiments have shown that at about 100 ms, the signal from the retina would have reached the prefrontal cortex. And between 180–260 ms, we can activate the motor of our hand for performing simple decision-making tasks.

Research shows that the prefrontal cortex has four areas that solve problems in a hierarchal sequence depending on the degree of complexity. The prefrontal cortex involves more top-down attention influence when there are more uncertainties (or noise data). Also, prior information affects decisions.

Chapter 4

MANY FACES OF MEMORIES — INVESTIGATING THE HUMAN MULTIPLE MEMORY SYSTEMS

Memory is essential to all learning processes (Ng, 2003). Whatever is learned, needs to be somehow stored and able to be retrieved or recalled at the right time for performing tasks, making decisions, thinking, or even just daydreaming. In the brain, the synapses are the key basis for information storage. Detailed studies by scientists have shown that there are multiple different memory systems in the brain.

Two broad ways of categorizing the multiple memory systems are based on information storage time and the type of information stored. For detailed studies of memory systems, you can refer to the following references: (Baddeley, 1992; Willingham, 2001; Eichenbaum and Cohen, 2001; Anderson *et al.*, 2004; Styles *et al.*, 2005; Kolb *et al.*, 2006).

This chapter discusses the following:

- The different types of memory systems based on information storage time and the type of information stored.
- What are the declarative and non-declarative memories?
- The role of the hippocampus in memory.
- Some interesting observations of the human memory.
- The human memory in chronological age.

Memory Systems Based on Information Storage Time

When categorizing memory systems based on operation time (or according to the effective time span that memory can be recalled), we can classify them as:

- Sensory Memory
- Working Memory (encompassing short-term memory)
- Long-Term Memory

Figure 4.1 shows a broad schematic diagram of memories that could potentially be mapped to a computational model for information processing purposes.

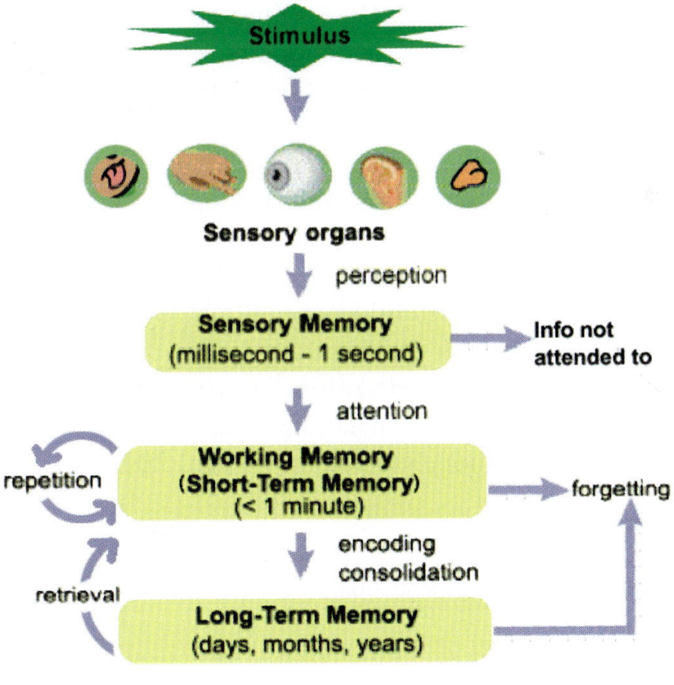

Fig. 4.1. A schematic diagram of memories. (Picture adapted from http://thebrain.mcgill.ca/)

Sensory Memory (SM). Sensory memory is memory from our immediate sensors, such as iconic memory[a] for visual stimuli, transsaccadic memory[b] for our eye movement, and echoic memory[c] for aural stimuli. Sensory memory helps to preserve accurate representation of the physical features of sensory stimuli for a few seconds or less. This brief timing is sufficiently long to give us a sense of continuity of the world without interfering with new sensory impressions, and in turn allows us to extend the availability of information acquired from the environment.

Working memory. Working memory holds information temporarily in the order of seconds to minutes. It holds some information and knowledge long enough for us to perform tasks, such as reasoning, thinking, and problem solving. In 1956, Miller famously proposed that the working memory recall is limited to about seven chunks or seven meaningful units of information (Miller, 1956). Research in 1985 indicated that when steps are taken to prevent the rehearsal or chunking of the presented items, the limit appears to be closer to about four units of information (Mandler, 1985). Further experiments also show that working memory capacity is dependant on one's ability to control attention (Kane *et al.*, 2004) and also the ability to inhibit irrelevant information (Gernsbacher, 1993; Lustig *et al.*, 2001). Professor Nelson Cowan, University of Missouri-Columbia, indicated that it is not one function of attention specifically, but attention more generally (such as zoom-in attention with a specific goal or to zoom out with a scope of attention, i.e. to apprehend the maximum number of targets) that is important for individual differences in working memory's capacity (Cowan, 2005).

---

[a] Iconic memory stores visual images for a very short time span, i.e. about half a second.

[b] Transsaccadic memory stores the position and orientation of our eye's saccadic movement.

[c] Echoic memory is the brief mental echo that persists after information has been heard.

According to Professor Alan Baddeley,[d] there are four components of working memory, namely the central executive system and three slave systems, i.e. the phonological loop, visuospatial sketchpad, and episodic buffer (Baddeley, 2000).

The phonological loop (PL) helps to hold and manipulate both verbal (speech related components) and auditory (sound related components) information. According to Willingham (Willingham, 2001), PL comprises of two parts, namely the phonological store and articulatory control process. The phonological loop can repeat about two seconds of auditory information, while the articulatory control process writes information into a phonological store. This is said to be the mechanism of "self-talk" — for more information see (Willingham, 2001). Other research on the phonological loop has found that its capacity is highly correlated with vocabulary in children. Baddeley *et al.* (1998) reviewed a number of studies investigating this correlation. The researchers found correlations between the digit span and vocabulary size, and between the ability to repeat long non-words and vocabulary size.

The visuospatial sketchpad (VSSP) has a similar function as a phonological loop, with the exception that this is exclusive for visual and spatial information only. It is a work-pad on which you can "see" things in your mental states (or mental picture). For example, this is where you will visualize a classroom scene or a picture of your family in a living room.

Baddeley indicated that information from the phonological loop and visuospatial sketchpad are coordinated by the central executive (CE). The CE also performs the role of controlling attention of information, i.e. CE will play an important role to divide mental resources to different parts of a task. The functions of CE include the initiation of retrieval of long-term memories, planning of future actions, usage of decision-making processes, and integration of new information.

---

[d] Alan Baddeley is a professor of psychology at the University of York. He is known for his work on working memory, in particular for his multiple components model.

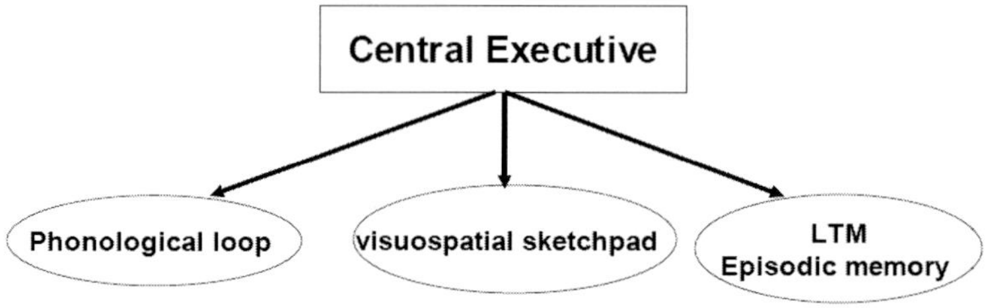

Fig. 4.2. A block diagram of working memory adapted from Baddeley's model (Baddeley, 2000).

In 2000, Baddeley added the episodic buffer (the third slave system) to link information across domains to form integrated units of visual, spatial, and verbal information with time sequencing or chronological ordering such as the memory of a story or a movie scene. The episodic buffer is also assumed to have links to other long-term memory (LTM).

Working memory encompasses the short-term memory (STM). Note: as the research of memories has become more refined, it has become increasingly apparent that the original concept of STM as a mere temporary storage for long-term memory is too simplistic. Other critics have also indicated that working memory is not well-defined and have come out with their own terminology, such as

- Conceptual short-term memory (CSTM coined by Mary Potter)
- Working short-term memory (WSTM)
- Working long-term memory (as opposed to WSTM)

The prefrontal cortex (PFC) is widely regarded as the place where WM and short-term storage of information are performed (Kolb *et al.*, 2006).

Remark: Perhaps at some course level, we can model the working memory. Modeling can include the detailed connection of CE, PL, VSSP and the association to LTM. We can also mimic the process of automation to free up the machine's working memory to do other tasks. Machines can perform repeated tasks automatically, i.e. sequential actions of the task retrieved directly from the

LTM. Note: when we remember new facts by repeating them, we are actually passing the information through the hippocampus several times to strengthen the associations, i.e. LTM build-up.

Long-term memory. Long-term memory (LTM) can be considered a warehouse of all experiences, events, information, emotions, skills, words, categories, rules, and judgments that have been attained from sensory and short-term memory (Gerrig and Zimbardo, 2005). Long-term memory is mainly processed in the temporal lobe, especially for verbal information. Information encoded into long-term memory after a long period of time would not require the intervention of the hippocampus as the various cortical regions are able to associate with one another to reconstruct memory. LTMs are also stored in various cortical regions.

It has been widely believed that the LTM is recorded in the neuronal circuits through alteration in the efficacy of existing synapses through long-term potentiation (LTP) and long-term depression (LTD) (discussed in Chapter 2). Synaptogenesis[e] and synapse elimination are alternative ways that neuronal circuits may be altered. Memories are retrieved from storage by chemical and electrical activity of neurons, which generates synaptic potentials determined by the pattern of synaptic connectivity between them.

Memory Systems Based on the Type of Information Stored

If we classify memory in terms of the types of information stored, LTM can be grouped into declarative (explicit) and non-declarative (implicit) memories. Declarative memory can be further divided into episodic and semantic memory, while non-declarative memory consists of both procedural memory and conditioning.

Declarative memory (also known as explicit memory), refers to the storage and retention of information with conscious effort, and the memory retrieved can be expressed in language. For example,

[e] Synaptogenesis is the formation of new synapses.

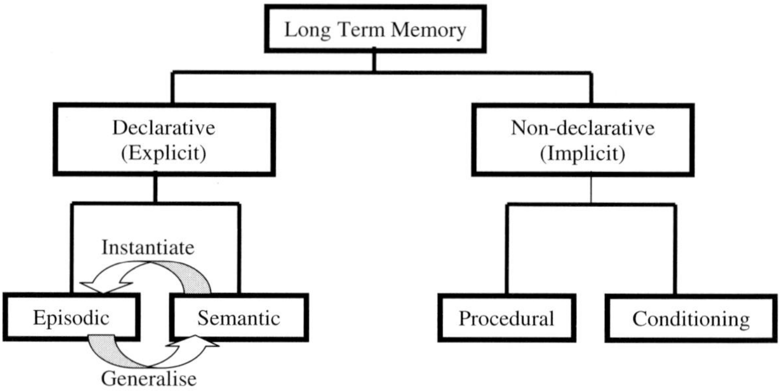

Fig. 4.3. An overview of long-term memory breaking down into declarative (explicit) and non-declarative (implicit) memories.

what is your friend's telephone number, when is your wife's birthday, and when is the last time you went for your medical check-up? Declarative memory can be further refined into episodic and semantic memory.

In contrast, non-declarative memory involves information that is acquired and retrieved at the unconscious level (without conscious effort). These are actions, habits, or skills learned through repetition and practice. For example, riding a bicycle, driving an automobile, skipping a rope, and playing badminton. It is interesting to note that long-term memory constitutes a large part of both declarative and non-declarative memories.

Damage to the medial temporal lobe (hippocampus is located in the medial temporal lobe) in the brain will impair declarative memory. Evidence for this has been obtained from amnesic patients. In medical history, a patient known as H.M., had a lesion in his medial temporal lobe. H.M. was studied by scientists with regard to his memory after the lesion. H.M. exhibited profound forgetfulness against a background of intact intellectual and perceptual functions. For example, if you have a conversation with him just a while, and you go to the toilet and return, he will not remember you had a conversation with him, let alone the conversation content.

No one knows exactly where the enormous LTM database is stored, but it is clear that the hippocampus is necessary to "file away" new declarative memory, and the cerebellum is likely to play a role in non-declarative memory.

Declarative Memory

Declarative memory can be further classified into semantic and episodic memory (Tulving, 1972). These two types of memories differ from each other by the kind of retrieval cues required to recall a particular event or fact.

Semantic Memory

Semantic memory refers to the generic, categorical memories such as meanings or words, what they refer to, facts and their associations, and rules and formulae that allow us to manipulate symbols. For example "an apple is a fruit". This kind of information can be made available without any reference to the time and place at which the event had occurred, such as with formulae or mathematical equations. Thus, when we recall facts, we do not have to think of the original learning episode in order to retrieve information. In other words, we still can obtain knowledge from forgotten experience.

When we recall semantic memory, we will activate the frontal and temporal cortices. The temporal lobe is known for the corresponding activation of the fact in question, while the frontal cortex is (speculated by Tulving) required for reaching consciousness.

Episodic Memory

Episodic memory helps to preserve individual, specific memories of personal experiences or episodes. For example, "What did I eat for breakfast today?" In order to recover episodic memory, one would need retrieval cues that specify either the time at which the episode or event occurred or the content of the events. For instance, the word "Paris" is needed for someone to recall incidents that happened during his trip to Paris. However, specific memory representations of events may or may not be formed depending on how the information is encoded. For example, one would not develop a specific memory of brushing teeth for the 10th or 11th time.

The hippocampus plays a fundamental role in episodic memory by reactivating a particular activity pattern in various regions of the cortex. After a period of continuous reactivating, the various cortical regions activated during recall will be strongly linked to one another. Therefore, the hippocampus is no longer required to act as a bridge between the various cortical regions after the strong link is established. As such, information in the long-term memory does not get "erased" even if one suffers a lesion in the hippocampus.

Autobiographical memory. Autobiographical memory is regarded as a special case of episodic memory. It is memory that captures self and personal involvement, such as one's emotions, thoughts, and self-reflection.

Interaction between semantic and episodic memory. It has to be noted that although semantic and episodic memory are different in some ways, they have to interact to allow the acquisition of new semantic knowledge (Tulving, 1972). There are times when we can forget our facts. One good way to aid in recall is to use additional retrieval cues, such as using original learning contexts. By transforming semantic memory into episodic memory, we may recall the necessary knowledge.

Tulving's view of the relationship between episodic, semantic, and procedural memory is one of increasing encapsulation, as episodic memory is said to be dependent on semantic memory, which is in turn dependent on the non-declarative's procedural memory. Non-declarative memory will be discussed in the next section.

Non-Declarative Memory

Implicit information is being encoded via a "data-driven" process, which means that information is passed from sensory receptors into the brain. In the brain, sensory information is processed in the subcortical and cortical regions. For example, visual information about an object from visual receptors is passed to the lateral geniculate nucleus (LGN) in the brain, followed by the occipital cortex, and finally, to the temporal lobe for object recognition.

Non-declarative memory can be further broken down to two types, namely the procedural memory and conditional memory

Procedural memory. Procedural memory enables us to perform tasks without conscious retrieval of information from our long-term memory store. In other ways, it is a procedure so standard that it is done automatically, such as riding a bicycle. As a result, it may be difficult for us to describe expert skills although we are able to perform well in those skills. It was found that patients with amnesia still have intact procedural memory and are also able to learn new skills. However, they cannot recall having learned that skill before.

Procedural memory therefore helps to increase the efficiency of performing a skill, otherwise it would be very tedious to go through every single step of activity to do a task. Procedural memory does not involve the hippocampus. Rather, procedural memory is associated with the cerebellum, basal ganglia, and the motor cortex, which are involved in motor control.

Conditioning memory. Conditioning memory is results from pairing events with parallel stimuli that may or may not be associated with the incident. In classical conditioning, a neutral event is paired with a response-evoking event, which eventually causes that neutral event to be able to evoke a response as well. For example, in fear conditioning experiments, a conditioned stimulus such as a tone is paired with an unconditioned stimulus such as an electric shock. After several trials of pairings, the tone elicits an autonomic response, resulting in fear-behavioral responses such as freezing. It is reported that the test showing the association between tone and shock is robust and long lasting. It has been suggested that conditioned fear associations help animals detect and avoid previously encountered threats throughout their lives (Quirk, 2002).

Role of Hippocampus in Memory

The hippocampus, in general, plays a critical role for declarative memory (both semantics and episodic memory). Literature reports that the hippocampus's roles include:

- To enable one to recall specific personal experiences, i.e. episodic types of memory (Vargha-Khadem *et al.*, 1997; Tulving and Markowitsch, 1998).
- To encode and retrieve sequential events (Fortin *et al.*, 2002).
- To mediate the capacity for memory recollection (Fortin *et al.*, 2004).
- To play an important part in the storing and processing of spatial information (Zola-Morgan and Squire, 1990) and navigation.

Some studies show that neurons in the hippocampus have spatial firing fields. These cells are called place cells. These cells fire when the animal finds itself in a particular location. The discovery of place cells suggests that the hippocampus might act as a cognitive map — a neural representation of the layout of the environment.

The medial temporal lobes (where the hippocampus and parahippocampal region are located) consolidate and integrate the memory. Information in the memory is then linked to the cortical association areas where the storage is more long-term. Cortical association areas are thought to perform higher level information processing. These areas are neither a function of motor nor sensory areas; but motor or sensory areas can be integrated to cortical association areas.

The hippocampus and related temporal lobe structures such as the entorhinal cortex, perirhinal cortex, and parahippocampal gyrus (Fig. 4.5) are responsible for the creation and consolidation of new declarative memories for storage elsewhere. In another word, the hippocampus and related structures are responsible for containing information in short-term memory, but are certainly not involved in storing long-term information. These structures have limited buffer capacity. The importance of the hippocampus in memory formation was discovered via experiments. And two famous case studies of

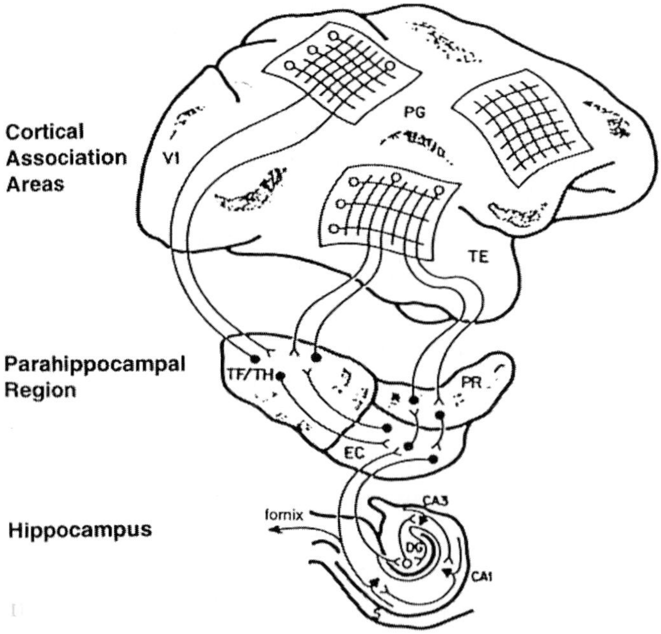

Fig. 4.4. The flow of information storage from the hippocampus to the parahippocampal region, and to the cortical association areas. (Diagram courtesy of Professor Patricia Bauer, Duke University.)

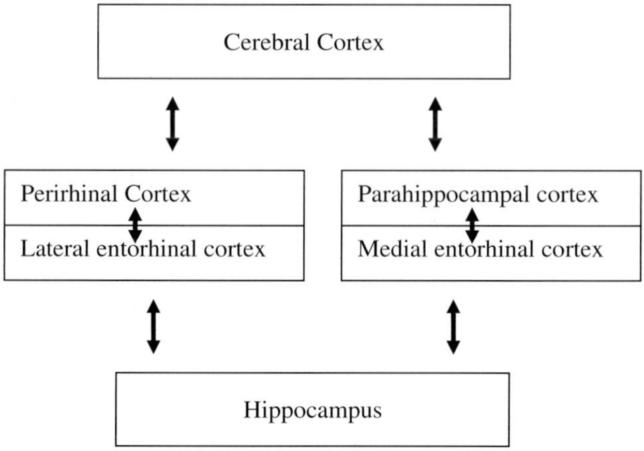

Fig. 4.5. A broad diagram of a medial temporal lobes organization.

amnesic patients, namely H.M. and R.B., contributed significantly to these experiments. In the case study of patient H.M., it was shown that hippocampus lesions are sufficient to cause anterograde amnesia. Anterograde amnesia is memory loss due to the inability to transfer information from short-term memory to long-term memory.

Another interesting observation is from the study of an Alzheimer's disease (AD) patient. The study shows that in AD patients, their hippocampal volume is generally less than in persons without AD. An AD patient in general exhibits the degeneration of hippocampal volume and regional cortical thickness. It is hypothesized that the white matter degeneration may result in a disconnection of cortical regions and contributes to cognitive dysfunction with aging and age-associated degenerative disease (Sowell *et al.*, 2004).

How the memory systems work in the medial temporal lobes is one of the key research areas.

What about non-declarative memory? For non-declarative memory, as of our understanding from current research, the basal ganglia, cerebellum, and premotor cortex play a more critical role than the hippocampus.

Let Us Sleep Over It

"*Memories of our life stories may be reinforced while we sleep*", MIT researchers reported on Dec 17 2006 in the advanced online edition of Nature Neuroscience.

A number of experimental studies indicate that sleep helps in reinforcing and consolidating memories. Memory consolidation is a process whereby memories are transferred from short to longer-term stores, and possibly reorganized into more efficient forms. A study shows that sleep replays awake experience in the cortex and hippocampus. However, whether temporally structured replay occurs in the cortex and whether the replay events in the two areas are related are unknown (Daoyun and Wilson, 2007; Louie and Wilson, 2001; Siapas and Wilson, 1998).

> "*We found that spiking patterns not only in the cortex but also in the hippocampus were organized into frames, defined as periods of*

stepwise increase in neuronal population activity. The multicell firing sequences evoked by awake experience were replayed during these frames in both regions. Furthermore, replay events in the sensory cortex and hippocampus were coordinated to reflect the same experience. These results imply simultaneous reactivation of coherent memory traces in the cortex and hippocampus during sleep that may contribute to or reflect the result of the memory consolidation process." (Daoyun and Wilson, 2007).

Interesting Observation of Human Memory

1. *Memory is associative.* Memory is associative, and thinking about one thing can get you thinking related thoughts. Because of memory's associative property, you can remember a new piece of information better if you can associate it with previously acquired knowledge that is already firmly anchored in your memory. And the more meaningful the association is to you personally, the more effectively it will help you to remember. So taking the time to choose a meaningful association can help you remember new information.

   Some people use this technique to increase their ability to remember more things. Tricks have also been derived to remember number sequences by using visual imaging as we associate visual imaging better, more so if we can associate the image with a story.

   Also, in your brain's memory systems, isolated pieces of information are memorized less effectively than those associated with existing knowledge. The more associations between the new information and things that you already know, the better you will learn it. For example, you will have an easier time remembering that the entorhinal cortex is connected to the hippocampus via the dentate gyrus if you already have some basic knowledge of brain anatomy.
2. *Memory is affected by attentiveness and concentration.* Memory is affected by the degree of attentiveness. Attentiveness is one of the critical tools that engrave information into memory.

Thus, attention and concentration deficits can radically reduce memory performance.

3. *Memory is strengthened more if it is interesting, motivated, and needed, or of necessity.* It is easier to learn when the subject matter fascinates you. Human memory systems have been shown to be enhanced when the motivation factor and interest are strong.
4. *Memory is affected by emotional state.* The more intense the emotions and mood, the higher the chance the event will be remembered. This also has led some scientists to believe in the existence of emotional memory.
5. *Memory can be retrieved via context cues.* While we are memorizing information, we also remember the contextual contents such as location and smell of the situation. Therefore, through a series of associations we are able to retrieve the recorded information from memory using contexts as retrieval cues.
6. *Some memories are never forgotten.* This is sometimes known as flashbulb memories. These refer to memories formed when some personal significant event occurs, and the whole scene is encoded into memory — for example, the first time you won a prize, your first date, a close friend of your passing away, or world events that strike you hard, such as the 9/11 World Trade Tower collapse. You may remember very clearly where you were or where you read and heard the news (Neisser and Harsch, 1992).
7. *Memory is vulnerable to "post-event information"*. Memories are vulnerable to "post-event information" (where post-event information involves facts, ideas, and suggestions that come along after the event has happened) (Loftus, 2003). You can, unknowingly, integrate this information into your memory, modifying what you believe you saw, heard, and experienced. Over time, you can integrate post-event information with information you gathered at the time of the event in such a way that you can't tell which details came from where, combining all this into one seamless memory. Some people find this finding disturbing as they would like their memory to be always accurate and reliable. It is true that our memory is accurate most of the time, but we also experience times when what we

remember is not totally true. Can you recall one of these incidents? Or an incident you know from your friend or colleague whose memory has "slipped" and the truth is not as they said they remembered? (Note this is assuming you have evidence to prove that what you remember is correct.)

8. *Memory constantly goes through reconstruction.* Whenever we want to recall (remember) or relate a story, we have to reconstruct it from elements scattered throughout various areas of our brains. Some scientists believe that there is a constant on-going process of reclassification resulting from continuous changes in synapses and possible neural pathways and parallel processing of information in our brains.
9. *Memory is not entirely faithful/trustworthy/reliable.* When you perceive an object (say a car), groups of neurons in different parts of your brain process the information about its shape, color, smell, sound, and so on. The brain draws information from different neurons in different parts of the cortex to form a perception of the object. Subsequently, whenever you want to remember the object, you must reconstruct these relationships. The parallel processing that your cortex does for this purpose can alter your memory of the object.

Human memory, in many circumstances, is very accurate. But research has demonstrated that memory can also be prone to distortion and is occasionally untrustworthy.

You can improve your memory capacity by making effort to repeat and integrate information by paying clear attention to how you want to integrate the information. One form of integrating information is to associate what you want to remember with information with which you already have a strong link. For example, to remember a password, you may like to associate it with a phrase you will recall easily.

"*Human beings feel attached to their remembered past, for the people, places, and events that we enshrine in memory give structure and definition to the person we think of as our "self." If we accept that memory spills over into dreams and imagination, then how do we know what's real and what's not?*" Loftus (in her research on why people feel uncomfortable). Loftus's research shows that memory is more prone to error than many people realize. Our memory system can be infused with compelling illusory memories of important events. These grand memory errors have contributed to injustices that could have been avoided or minimized (Loftus 2003b).

Human Memory in Chronological Age

"*Memory is born anew every day*" E. Galeano

People in general cannot remember specific events from the early years of their lives. Studies have shown that humans can recalled very few memories before the age of three. And if one would to recall their entire lifespan, the amount one can recall before age seven tends to fall off sharply in comparison to other periods (Fig. 4.6).

One of the key points shown in Fig. 4.6 is that we start to remember more of our life stories after the age of six or seven. Picture taken with permission from Professor Patricia J. Bauer (Duke University) (Talk in BU/Department of Psychology by Professor Patricia J. Bauer: "Getting a life: Contributions from developments in episodic and autobiographical memory" on March 2007).

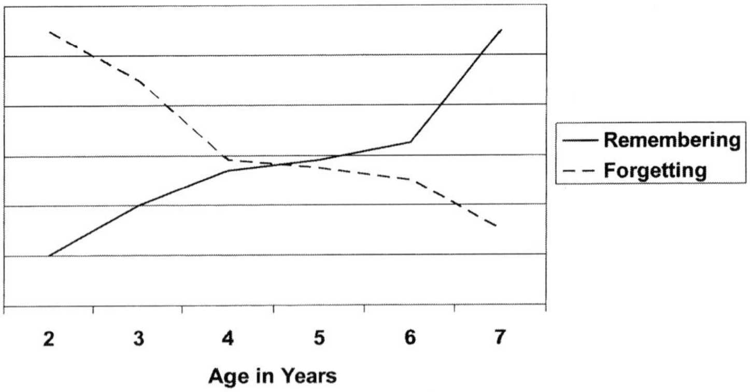

Fig. 4.6. A chart of the rate of remembering and forgetting with respect to age.

Why is the adult distribution of autobiographical memories seen around six to seven years old, but not before? It is interesting to know that studies have shown that the overall size of the brain is relatively mature (about 95% of the adult volume) at the age of six to seven years old. There seems to be a correlation between the brain growth and the ability to remember. When the growth is stabilized at six to seven years old, our ability to remember also increases. Later studies have also shown that after the age of six to seven years old, the brain structure continue to change but at a very local and discrete pattern (Sowell *et al.*, 2004). One possible hypothesis on why young children less than six years old have amnesia is due to their growing brain, which is undergoing constant changes and recalibration of their information storage location, and this makes the brain forget the information. Hence, information could be lost or "forgotten", and after seven years old, when the brain stop growing, information storage becomes more stable and we can remember better.

Summary

The human brain has many different memory systems. This chapter covers the different types of memory systems based on information storage time and the type of information stored. Memory systems based on operation time can be classified into sensory memory, working memory (including short-term memory), and long-term memory.

Long-term memory can be further categorized into the type of information stored. These are namely the declarative memory (also known as explicit memory) and the non-declarative memory (or implicit memory). Declarative memory stores information with conscious effort and the memory retrieved can be expressed in language. Declarative memory can be further refined into episodic and semantic memories. In contrast, non-declarative memory involves information that is acquired and retrieved at the unconscious level (without conscious effort). These are actions, habits, or skills learned through repetition and practice. Non-declarative memory can be divided into procedural and conditional memories.

The hippocampus plays a critical role in declarative memory. These include recall-specific personal experiences (episodic memory), sequential encoding and retrieval of events. The medial temporal lobes, where the hippocampus is located, are critical for the consolidation and integration of memory. Lesions or damage to the medial temporal lobes, particularly at the hippocampus, will cause anterograde amnesia. Research has shown that in Alzheimer disease patients, their hippocampal volume is generally less than for persons without Alzheimer disease. It is hypothesized that the white matter degeneration may result in a disconnection of cortical regions and contributes to cognitive dysfunction with aging and age-associated degenerative diseases.

Many other interesting observations of human brain memory systems are documented. These include: memory is vulnerable to "post-event information" and memory constantly undergoes consolidation and reconstruction. The human brain seems to be stabilized[f] at the age of six- to seven-years old, and our ability to remember increases after this age.

---

[f] Stabilized here refer to maturity of brain growth and the increase in the ability to remember.

Chapter 5

LEARNING LIKE A HUMAN: HOW DOES LEARNING TAKE PLACE IN OUR BRAIN?

How does the brain learn? How does learning occur in the brain's neural networks? How do we learn to make decisions? We learn at almost every instant; our synapses change constantly as we see and experience new events. Learning can be considered as the encoding of experience into memory. Without learning, memory cannot exist; without memory, learning is useless (Ng, 2003). One result of learning is the formation of different memories such as episodic, semantic, and procedural memories.

Memory, or more specifically long-term memory, is the record left by a learning process. Learning is needed for many reasons, and two key reasons from the perspective of building intelligent systems are:

- Building a "commonsense knowledge database" in long-term memory, and
- Continuously refining and aligning the "commonsense knowledge databases" during everyday cognitive activities and interaction with the environment.

This chapter discusses learning and its importance for intelligent systems as follows:

- Discusses how humans learn;

- Presents how plasticity and stability give rise to learning;
- Discusses the various classes of human learning;
- Including perceptual, stimulus-response, motor, and relational learning;
- Examines the role of dopamine in learning, and
- Addresses cognitive learning.

How Do Humans Learn?

According to (Ormrod, 1999), the definitions of learning vary with the perspective of the theorist:

- From a behaviorist perspective, learning is a relatively permanent change in *behavior* due to experience. This refers to a change in behavior, that is, an external change that we can observe.
- From a cognitive perspective, learning is a relatively permanent change in *mental associations* due to experience. This definition focuses on a change in mental associations, that is, an internal change that we cannot observe.

From a neuroscientific perspective, learning involves changes in the nervous system that occur as a result of new sensory experiences. These changes in the nervous system include physical changes in the size, shape, and flow of the neurotransmitter at synapses as well as new growth of neurons. These alterations of the brain at the neural level systematically affect and shape our behaviors. This ability to adapt and change is known as neural plasticity. Note that neural plasticity means the neurons are able to alter and reshape themselves so as to grow and form new connections.

Few of the details and mechanisms behind neural plasticity are known. The best-known processes are long-term potentiation (LTP) and its opposing process, long-term depression (LTD). These processes are discussed in Chapter 3. LTP/LTD involves the increased/decreased conductivity of synaptic connections between neurons. LTP/LTD occurs as a consequence of increased synchrony/asynchrony in the activity of pre-synaptic (sending) and post-synaptic (receiving) neurons. LTP/LTD is presumed to produce learning by differentially facilitating the associations between stimuli and responses.

Besides LTP/LTD, there are two alternative proposals regarding the mechanism of neural plasticity; these are, namely, the *in-vitro* reinforcement (IVR) (Stein and Belluzzi, 1989; Stein 1994, 1997) and the hedonistic synapses (HS) (Seung, 2003). Some details of these proposals are presented in the note for Chapter 5.

How Plasticity and Stability Give Rise to Learning

How does the brain learn in the first place? How can the adaptability of neurons in the brain give rise to learning? How do we learn new things while retaining what we have already been learned?

The adult brain. The healthy adult brain is both plastic and stable, i.e. it is adaptable and robust in its ability to retain information without suffering catastrophic chaos or instability. Recent research has shown that even in the mature adult brain, stem cells continue to be produced. Researchers from John Hopkins University (USA) and National Cheng Kung University (Taiwan) have found that new adult neurons (neurons newly generated in the hippocampus of an adult) show a pattern of changing plasticity very similar to that seen in the brain cells of newborn animals. These new adult brain cells have a "critical period" in which they are highly plastic before they differentiate into less plastic, mature brain cells. In newborn animals, such a critical period enables an important, early burst of wiring of new brain circuitry with experience. This finding suggests that new adult neurons are equipped to facilitate new learning through experience-dependent wiring of new connections and pathways (Ge *et al.*, 2007).

This same group of researchers also discovered that at the molecular level, plasticity depends upon neurotransmitter receptors being functionally intact. A neurotransmitter is a chemical-liquid signal secreted by a presynaptic neuron onto a neighboring, postsynaptic neuron that triggers a nerve impulse in the receiving neuron. Neurotransmitter receptors mediate the ability of new neurons to undergo LTP. Further experiments revealed that the ability of a new neuron to undergo LTP in response to theta burst stimulation depended on the cell's age. Subtle alterations in receptor populations are the means by which the brain wires the preferred pathways in the process of learning and memory. The researchers concluded that adult neurogenesis (the process of creating new neurons) may represent not merely a replacement mechanism for lost neurons, but an ongoing developmental process that continuously rejuvenates the mature nervous system by offering expanded capacity of plasticity in response to experience throughout life. In other words, new adult brain cells may give the adult brain the same kind of learning ability that young brains have while still allowing the existing, mature circuitry to maintain stability. Note that neurogenesis in the adult brain and LTP/LTD are complementary, not competing processes by which learning can take place.

The adolescent brain. A research group lead by Professor Toga Arthur and colleagues from the University of California (Los Angeles) discovered that the brain continues to undergo surprisingly dramatic anatomical changes between the ages of three and 15 years old. More specifically, their research indicates that from ages three to six, the most rapid growth takes place in frontal-lobe areas involved in maintaining attention to tasks and planning or organizing new actions. By contrast, during the period from six to puberty, the spike in gray-matter growth shifts to the temporal and parietal lobes, which play a major role in language skills and spatial relations. Beyond puberty, the growth rate of these brain areas falls off fast. This may explain why, as a rule, the ability to learn languages declines sharply after the age of 12 (Suplee, 2000).

Further research from this lab indicates that as children age, the growth of gray matter occurs in a wave, moving from the front of the brain to the rear. This finding relates to an interesting observation made by Dr Jay Giedd, from a child psychiatrist with the

National Institute of Mental Health in Bethesda (Suplee, 2000): "*In the womb and during the first 18 months of life" when the brain undergoes its most drastic changes, "an infant does not have much say about the way things turn out. But during the teenage years, "a person has a lot to say" about the way his brain develops.*"

Their findings indicate that the teenage years are a kind of critical period during which the brain is optimized for commonly practiced abilities. This means when a person does sports or academics or music as a teenager, the circuits that determine these abilities are going to be hard-wired as the brain matures.

The infant brain. The amount of plasticity in an infant's brain is remarkable. Studies using adult monkeys have shown that hippocampal lesions impair spatial relational learning, i.e. the ability to learn and remember a particular location in relation to the environmental spatial cues (see previous sections for a discussion of the research showing that hippocampal lesions disrupt the formation of new declarative memories in humans). David Amaral, a professor at the University of California (Davis) and research director of the MIND Institute, has shown that, unlike adult monkeys, hippocampal lesions in infant monkeys do not disrupt the learning of spatial relations. The preservation of cognitive function in infant monkeys following hippocampal lesions spotlights the high degree of plasticity found in the infant brain. However, where this plasticity takes place and precisely how it reorganizes this circuitry is currently unknown (Lavenex *et al.*, 2006). It is interesting to note that the newborn brain has many times more neurons and synapses than an adult brain and that many of these neurons and synapses may not yet be fixed on a particular function or circuit during this early stage of development (note genetically predefined functions are discussed in earlier chapters).

Learning to use vision after the critical period. Another breakthrough in our understanding of the brain and plasticity was obtained by researchers studying the patient SK. SK was born with congenital aphakia (a condition where the eyeball develops without a lens). In principle, he was blind and could see only after he was given a special pair of glasses in July of 2004 (under the project Prakash led by Professor Pawan Sinha, a neuroscientist at MIT). SK had passed the so-called "critical period" of visual learning. This "critical period" in visual learning occurs between the ages of one and three years. Previous studies led many scientists to believe that if a person does not learn to see during this "critical period", then the person will never see normally. But 18 months after getting his glasses, SK surprised everyone. He had begun to make sense of his world; he recognized more complex objects with varying colors and brightness. This demonstrated the plasticity and adaptability of the brain to learning (Madavilli, 2006.)

Forms of Human Learning

How do our brains select what to learn? When does learning take place? There are different possible forms of human learning discussed by researchers from different fields. Most of these forms of learning overlap or are related to one another. Some of these forms of human learning include

- Perceptual learning, which functions to identify and categorize objects and situations. It involves changes within the sensory systems of the brain.
- Stimulus-response learning involves making a response when a particular stimulus is present, i.e. the ability to learn to perform a particular behavior when a certain stimulus is present. It is the establishment of connections between perceptual and motor systems.
- Motor learning involves the change in motor systems in the brain. Motor learning is the process by which we learn to acquire precise, coordinated movements, like those that are required for running, serving a tennis ball, etc. It needs sensory input, and hence it involves its own stimulus-response learning mechanism.
- Relational learning involves identifying connections between stimuli through the connections between different areas of the association cortex.

Through the study of these four forms of human learning, we hope to understand and gain some insight into the mechanisms of human learning.

Perceptual Learning

The aspects of perceptual learning discussed here include visual learning and auditory learning. This form of learning appears to take place in low level cortical areas such as V1 (in the visual cortex) or A1 (in the auditory cortex) and sensory association cortex. Higher level cortical areas receive sensory information from other cortical areas while lower level cortical areas receive sensory information from sources nearer the sense organs, outside the cortex.

Visual perceptual learning. It is known that the visual pathway consists of two streams, namely, the ventral (what) stream and the dorsal (where) stream. The ventral stream carries information as to "what" is the object. Results from monkey experiments have shown that the ventral stream's ability to differentiate visual patterns requires intact connections between the visual cortex and inferior temporal cortex. The dorsal stream carries information as to "where" an object is in space, such as its orientation, direction, and position.

Perceptual learning in the lower visual cortex has been studied extensively by Professor Takeo Wanatabe and Dr Aaron Seitz in Boston University, Department of Psychology. A subsequent section describes some of their latest findings.

Auditory perceptual learning. The auditory system is capable of temporal processing across a wide range of scales from microsecond timing (up to 20 KHz) to several seconds (20 Hz). In the auditory system, this range is used in the ability to discriminate the order, interval, and duration of sound, which is important for speech processing (Shannon *et al.*, 1995). Deficits in this range of temporal processing may contribute to some language-based learning disabilities (Merzenich *et al.*, 1996). However, even for simple interval discrimination tasks, little is known about the neural mechanisms and areas involved in timing.

(Karmarkar and Buonomano, 2003) conducted an experiment to study whether learning on an auditory interval discrimination task generalizes across stimulus types, intervals, and frequencies. The degree to which improvements in timing carry over to different stimulus features constrains the neural mechanisms underlying timing. Human subjects trained on a 100–200 msec interval discrimination task showed an improvement in temporal resolution. This learning generalized to a perceptually distinct duration stimulus, as well as to the trained interval presented with tones at untrained spectral frequencies. Their results indicate that the brain uses circuits that are dedicated to specific time spans and that each circuit processes stimuli across non-temporal stimulus features. The patterns of generalization additionally indicate that temporal learning does not rely on changes in early, subcortical processing, because the non-temporal features are encoded by different channels at early stages.

Visual perceptual learning from Professor Takeo Watanabe and Dr Aaron Seitz. Perceptual learning here refers to improvements in sensory abilities the at low level visual cortex, such as V1 (Seitz and Watanabe, 2005). In Seitz and Watanabes' experiment, they asked subjects to focus on the center of the screen at a small circle with a letter in it. This circle forms the focus of attention of the subjects (see figure below).

Fig. 5.1(a) shows the procedure of the exposure stage in Seitz and Watanabes' experiment. A 5% coherent motion direction (shown as up in the figure) was temporally paired with the target letters (shown as white); three other directions were paired with distractor letters (shown as black). Fig. 5.1(b) shows a change in performance between the pretest and posttest. A significant improvement was found for the direction temporally paired with the targets when assayed at 10% motion coherence. No significant changes were found for 5% coherent motion (Seitz and Watanabe, 2005).

The task given to the subject was to remember which letters were in white color (only two letters were in white, with the rest in black). Surrounding the small circle were numerous dots (these dots are task-irrelevant since the dots have nothing to do with

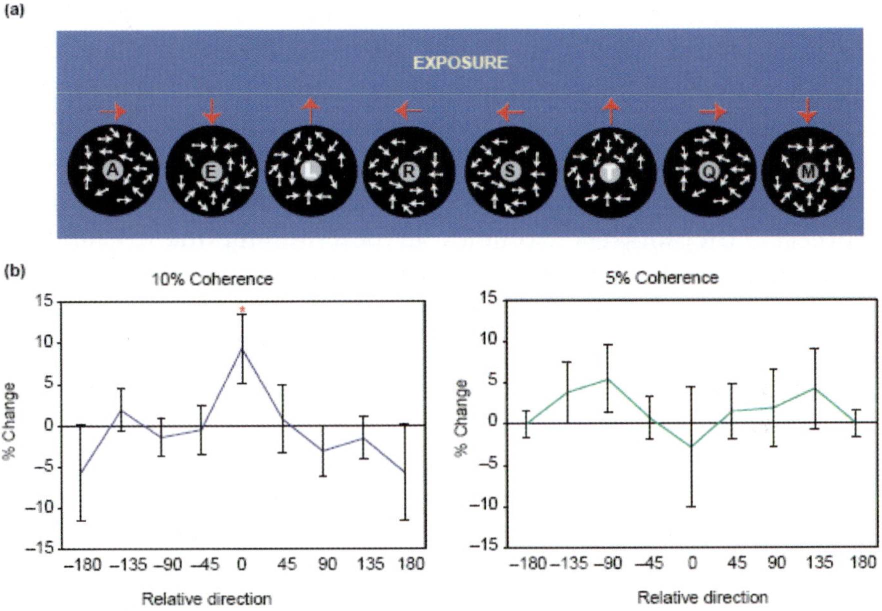

Fig. 5.1. Procedure of exposure stage in Seitz and Watanabes' experiment.

whether or not the subject recalls the letters in white). A percentage of these dots will move in the same direction with the same speed. The rest of the dots will move randomly. 5% correlation is the term used to describe that 5% of the dots move together in the same direction with the same speed. Through exposure of these dots, the subject experiences task-irrelevant learning in that they will be able to determine the direction of the correlated motion of the dots significantly better than chance level, even though they were focusing their attention to the task of recalling the white letters. At 5% correlation level, the motion of 5% of the dots is not obvious to the subject. A version of the experiment setup is as follows: A pre-exposure test is given to subjects (to determine baseline accuracy rate). In this test phase, the correlation level is 10%. The number of trials in the test period is small, so this prevents the subject from learning the trials in the test phase. This is followed by an exposure phase.

In this phase, the subject is exposed to trials with 5% correlation level but the subject is not tested. After the exposure phase, the subject carries on with the second test phase. In this test phase, the correlation level is 10%. It was found that through the exposure phase with 5% correlation level (recall that at 5% correlation level, the movement direction of the dots is not obvious to the subject), the subjects' accuracy in determining the correlated movement of the dots increased. This indicates that there is **passive perceptual learning** — the subjects learn the correlated movement of the dots even though the dots are not the focus of attention. (Note: If the correlation level for both test phases were 5%, the increased change in accuracy via learning is too small to be noticed).

In another version of the experiment, the direction of the correlated movement of the dots is the same for all white letters (targets). The directions of the dots for black letters (distracters) are equally likely to be three out of the four possible directions (north, south, east, west). It is found that subjects only learn the direction of the dots for white letters. This suggests **associative perceptual learning** in that learning only occurs with targets. The direction of movement of the dots is not learned with non-targets/distracters. Hence, **perceptual learning** is associative and related to the conditional learning.

Seitz and Watanabe trained the subjects for 45 days and it was found that the task-irrelevant learned skill was retained for a long time (six months). Perceptual learning occurs in the low level visual cortex region (V1) and not in the sensory memory level. This enables long-term encoding of the task-irrelevant learned skill.

Figure 5.2 shows an updated proposed unified model framework of perceptual task-relevant and task irrelevant learning by Seitz and Watanabe. Reference (Seitz and Watanabe, 2005) and Watanabe's presentation in BU (March 30 2007, The Brain, Behavior and Cognition Spring 2007 Colloquia Series). This updated unified model shows the inhibiting control link from task target orienting to the irrelevant feature (courtesy of Professor Takeo Watanabe).

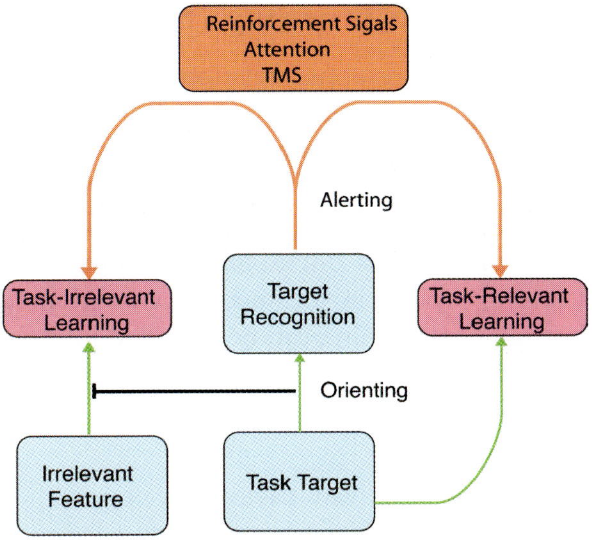

Fig. 5.2. A proposed unified model framework for perceptual learning.

Seitz and Watanabes' experiments support the model that both task irrelevant and task relevant learning occurs when task target and reinforcement signals interact at an appropriate timing. They also found out that through fMRI scans, DLPFC has a higher activity when the correlation level of the movement of the dots is increased (i.e. the DLPFC "notices" the correlated movement of the dots). When the correlation level of the movement of dots is increased, the DLPFC's activity increases but the activity of MT decreases. Therefore, they hypothesize that there is some inhibitory control via DLPFC to MT. This causes the subject to have a lower accuracy as they increased the correlation level of the movement of the dots. For example, at 5% correlation level, the DLPFC does not "notice" the correlated movement. There is little inhibition of DLPFC to MT activity, so the subject has a high performance accuracy. At 20% correlation level, DLPFC "notices" the correlated movement, inhibits the MT activity, and hence, the subject's accuracy level drops.

Stimulus-Response Learning (S-R Learning)

Learning stimulus and response relationships, or S-R learning, establishes connections between sensory and motor systems. A few related types of S-R learning such as conditional learning and operant learning will be discussed in this section. The computational techniques based on S-R learning are also discussed.

Conditional learning. As discussed in Chapter 4 on memory, there is a learning process in forming the conditioning memory. Learning through conditioning can be classified into classical conditioning (pioneered by Russian physiologist Ivan Pavlov) and operant conditioning (or instrumental learning).

- *Classical conditioning*: Ivan Pavlov developed the theory of classical conditioning through the study of dogs. From his perspective, learning begins with a stimulus response connection. Classical conditioning associates between two forms of stimuli, i.e. the unconditioned stimulus and the conditioned stimulus. The unconditioned stimulus (US) (e.g. a bowl of meat) that normally causes or provokes a response called unconditioned response (UR) (e.g. the dog salivates). And this is repeatedly associated with a conditioned stimulus (CS) (e.g. a bell — in normal cases the bell is neutral and does not provoke the response) will eventually cause the unconditioned response (the dog salivates) without any need for the unconditioned stimulus (the bowl of meat). Hence, classical conditioning learning is learning to pair the CS and US over many trials till a CS alone will produce a CR. Note that before conditioning, the bell (CS) should attract the dog's attention or elicit the orienting response to be conditioned.

  All forms of learning can be reduced to a conditioning phenomena according to Pavlov. Pavlov showed that conditioned responses originated in the cerebral cortex. Pavlovian conditioning mechanisms are said to be responsible for the formation of associations between stimuli and unconditioned stimuli. These associations can evoke an associated drive (hunger or sex), affective (pain or pleasure), or autonomic responses (Hall, 2001).

The classical conditioning is closely related to a computational temporal difference learning (TD) algorithm. TD learning will be discussed later in this chapter.

- Operant conditioning (or instrumental conditioning or instrumental learning) was pioneered and extensively studied by Edward L. Thorndike (1874–1949) by observing the behavior of cats trying to escape from home-made puzzle boxes. Operant conditioning involves an association between a response and a stimulus; allows an organism to adjust its behavior according to the consequences of that behavior. In Thorndike's cat experiment it was observed that cats took a long time to escape. With experience, ineffective responses occurred less frequently and successful responses occurred more frequently, enabling the cats to escape in less time over successive trails. Therefore, operant conditioning is the use of consequences to modify the occurrence and to form behavior. Unlike classical conditioning, operant conditioning deals with the modification of voluntary behavior through the use of consequences. These could be either:

  - "favorable" consequences (reinforcing stimuli), hence more likely to occur in future, or
  - "unfavorable" consequences (punishing stimuli), hence less unlikely to occur in the future.

The reinforcing or punishing stimulus in operant conditioning influences the amount of dopamine released and leads to synaptic changes. It then strengthens weak synapses (reinforcing) or weakens strong synapses (punishing). The reinforcing and punishing stimulus in operant conditioning is closely related to reinforcement learning. However, the operant conditioning learning in animals involves much more complicated behaviors than the computational model of reinforcement learning (Touretzky and Saksida, 1997). Reinforcement learning will be discussed in the later section.

Biologically, the following areas are involved in S-R learning:

Ventral tegmentum area (VTA). It is claimed that reinforcing stimuli activate neurons here, which stimulate the release of dopamine. The VTA is part of the midbrain and is rich in dopamine and serotonin neurons. The VTA is also related to two major dopamine pathways, namely:

- The mesolimbic pathway — which connects the VTA to the nucleus accumbens
- The mesocortical pathway — which connects the VTA to cortical areas in the frontal lobes.

The amygdala is involved in the detection of CS for reinforcement — if monkeys are trained that food follows a visual stimulus, then after their amygdala is lesioned, the monkeys forget the association

Lateral hypothalamus (LH) neurons become active when monkeys see food, but only when hungry — neurons show sensory-specific satiety; activity related to presence of reinforcing stimuli.

The prefrontal cortex secretes excitatory glutamate, which triggers bursts of dopamine to be released from neurons in the ventral tegmental area into the nucleus accumbens; this may serve as monitor for reinforcement-seeking activity.

Computational Techniques Based on S-R Learning

Temporal difference learning. Temporal difference (TD) learning is a prediction method. It is a combination of ideas from Monte Carlo techniques and dynamic programming (DP). It resembles a Monte Carlo method because it learns by sampling the environment according to some policy. And it is related to DP techniques because it approximates its current estimate based on previously learned estimates (a process known as bootstrapping).

The TD learning algorithm was originally conceived in part as a theory of Pavlovian conditioning (Barto and Sutton, 1982), and more recently, with connection to reinforcement learning and machinery in the brain. (TD learning is one way to explain how the brain performs learning).

Reinforcement learning. Reinforcement learning (RL) is by reward (positive) and penalty (or negative or punishment) methodology to teach a system to learn. The learning is due to adjustment of its weight dynamically in each iteration with presentation of the reward or penalty stimulus coming from the environment. RL is sometimes considered a special class of supervised learning. Note that RL is sometimes considered to be learning with a critic as opposed to learning with a teacher in supervised learning. RL theory remains one of the competing techniques in the study of how the neural mechanism learns for decision-making (Sutton and Barto, 1998; Dayan and Balleine, 2002; Camerer, 2003). More recently, the reinforcement learning process is shown to adjust representations of competing decision options, and hence, is used for prediction of future decisions (Cohen and Ranganath, 2007).

Associative learning. S-R learning (and so is motor learning) can be considered a special class of associative learning. Associative learning involves the ability to learn to associate events (and this includes the stimulus and response type of events) and the relationship between such events. New knowledge can be built using existing knowledge in an associative learning process. The stronger the associative link of the new knowledge to the preexisting knowledge, the more permanent the storage (or more effective the encoding of information in neurons) of the new knowledge will be. Associative learning can also be applied to new ideas and experiences. In psychology, new ideas and experiences can reinforce one another and are associated with enhancing the learning process. Related to associative learning are conditional learning and perceptual learning.

Motor Learning

Motor learning is the process of improving the smoothness and accuracy of movements. This includes learning new motor skills.

Motor learning is essential for complicated movements such as eating, speaking, playing (table tennis or an instrument), and climbing (stairs or a tree). It is also important for calibrating simple movements like reflexes, as parameters of the body and environment change over time. The cerebellum and basal ganglia are critical for motor learning.

Motor learning requires the following:

- Motor commands and the sensory feedback from the movement, either as raw data or its second order derivative data, such as the error signals, need to be stored in memory for learning to occur.
- Cognition is needed for motor learning. For example, a hockey goalie tries to predict the direction of a shot by searching for perceptual clues that provide advance information. A golfer without a clear shot to the green tries to remember how to hit a controlled fade. A figure skater about to perform a triple axle jump followed immediately by a triple toe loop must prepare for this combined action with the flexibility in mind to change the plan if something goes wrong. All of these are examples of decision-making processes regarding the anticipation, planning, regulation, and interpretation of motor performance. However, some motor responses are reactive and may not need deliberate cognitive processes.

Procedural learning and sequence learning (learning a sequence of movements) are part of motor learning (sometimes also known as sensory-motor skills), and these are mainly task-oriented activities such as typing, operating musical instruments, cleaning a utensil, assembling a bookshelf or learning the procedure to drive a car. These forms of learning normally involve feedback to correct the motor behavior and selective attention to determine what actions are appropriate. Hence, it has some form of supervised or guided learning in the early stage of learning the procedure. Once the skill or tasks are learned, the process becomes more automated in the brain and attention is given to other more important events such as predicting the motion due to variation in new environment, e.g., predicting the next limb moment in uneven terrain not encountered before.

Relational Learning

Relational learning involves identifying connections between stimuli through the connections between different areas of the association cortex.

- Spatial learning is learning about where you and others are in your environment. The hippocampus is involved in spatial learning. An experiment has shown that London taxi drivers have more hippocampal grey matter compared to ordinary folk (McGuire *et al.*, 2003). (Evidence of activation of the hippocampus during memory-driven spatial navigation (McGuire *et al.*, 1997). Evidence of the hippocampus's role in spatial learning from lesion studies (O'Keefe, 1993).)
- Episodic learning is remembering the sequences of events that occur as we see them.
- Observational learning is learning by watching and imitating other people.

Learning by imitation. Learning through imitation is related to observational learning. Learning by imitation seems to be the most basic for all humans, particularly young children. This form of learning was further supported by the discovery of mirror neurons. Mirror neurons are discovered by Giaccamo Rizzollati of the University of Parma (Italy). Rizzollati discovered that at the ventral premotor area of monkey frontal lobes, there are mirror neurons. Mirror neurons allow us to imitate the movements of others. Some scientists believe that mirror neurons play a key role in early infant imitation of action. These neurons show activity in relation both to specific actions one performs and matching actions others perform (Gallese and Goldman, 1998; Iacoboni *et al.*, 1999; Williams *et al.*, 2001).

All the different types of human learning that I have listed are interrelated to one another and they are also related to our higher thinking and reasoning.

Dopamine's Role in Learning

Dopamine is a neurotransmitter in the human nervous system. It plays a key role in learning, in particular, the following three key areas in learning:

- *Dopamine related to reward.* Professor Richard Depue, Cornell Unversity, indicated that the higher the level of dopamine, or the more responsive the brain is to dopamine, the more likely a person is to be sensitive to incentives and rewards. "*When our dopamine system is activated, we are more positive, excited and eager to go after goals or rewards, such as food, sex, money, education or professional achievements.*" — Richard Depue. He also showed that dopamine is strongly related to how well the prefrontal cortex holds information. "*To hold in short-term memory a spatial map of the environment, for example, you must have the dopamine system activated; without it, you cannot do this type of cognitive functioning.*" — Richard Depue (Lang, 1996).
- *Dopamine related to making choices.* A group of researchers at the University of Colorado Boulder studying Parkinson's disease patients reported strong evidence that dopamine in the brain plays a key role in how people implicitly learn to make choices that lead to good outcomes while avoiding bad ones. "*Often people will get a "gut feeling" that allows them to make a choice depending on how often it was associated with positive outcomes in the past. But people with Parkinson's disease often have difficulty making these kinds of choices.*" — Michael Frank. (Frank *et al.*, 2004).
- *Dopamine related to conditional stimulus.* The release of dopamine is observed to relate to the onset of a conditional stimulus. It has also been observed that the dopamine neurons are depressed when the expected reward is omitted. Thus, it is

reported that dopamine seems to encode the prediction error of rewarding outcomes. In nature, we learn to repeat behaviors that lead to maximized rewards. Dopamine is therefore believed by many to provide a teaching signal to parts of the brain responsible for acquiring new behavior (Schultz *et al.*, 1997).

Model of dopamine system for learning. As dopamine plays a key role in learning, modeling the dopamine system has become a key research focus. Some believe that the dopamine system should be modeled using a temporal difference learning mechanism (Schultz *et al.*, 1997; Daw, 2003). Temporal difference learning provides a computational model describing how the prediction error of dopamine neurons can be used as a teaching signal (Schultz *et al.*, 1997).

Others model the dopamine system as a positive and negative reinforcement learning mechanism. For example, (Frank *et al.*, 2004) modeled the dopamine system of the basal ganglia using reinforcement learning, which he called "carrot and stick" learning. In BU/CNS, (Brown *et al.*, 1999) modeled the dopamine system based on conditional learning. For a brief description of this model, refer to the note in Chapter 5.

Cognitive Learning

This section describes some other forms of cognitive learning. The term "cognition" comes from the Latin word for "knowledge" or "thinking". Hence, cognitive learning refers to more complex behaviors of human learning. These include affective learning, supervised/unsupervised learning, transfer learning and match/mismatch learning, among others. Humans start this learning process early in life and continue with it for their entire lives.

Affective learning. In human behavior studies, it is observed that humans do not separate emotions from cognitions, and from the educational learning perspective, learning is incomplete without attention given to the individual's interest, motivation, appreciation, and attitudes (Chickering, 2006; Owen-Smith, 2004). Research has demonstrated that emotion (such as fear, anger, and joy) affects learning. For example, a slight positive mood not only makes you feel a little better but also induces a different kind of thinking characterized by a tendency toward greater creativity and flexibility in problem solving, as well as more efficiency and thoroughness in decision-making. These effects have been found in many different ages and professions (Isen, 2000) (Picard *et al.*, 2004). We could also experience that in our everyday lives. Some believe that affective learning could be one of the early learning processes of an infant.

However, there is still very little understanding as to which emotions are most important in learning and how they influence learning. In order for learning to become truly rooted, a person has to have a deep emotional attachment to the subject area. Affective learning is to foster a love for learning before enforcing a strict learning strategy.

Here, we focus the affective learning on the cognitive and information processing aspects rather than the affective learning in pedagogy terms. How should affective learning be incorporated into the current machine learning techniques? How should machine-based rules be encoded and studied with learning interaction? Affective learning techniques include the OCC (Ortony, Clore and Collins' model, 1988; Ortony *et al.*, 1988) and Cathexis model (Velasquez and Maes, 1997) of emotion. Affective learning is in its infancy stage of research. MIT media lab is actively pursuing this research.

Supervised and unsupervised learning. Supervised learning needs a teacher/supervisor or teaching signals to adjust or correct its generated output given the input. Unsupervised learning does not need a supervisor or teaching signals. Biologically, there are many related examples for supervised and unsupervised learning in the brain. For example, the photoreceptors in the eyes are constantly changing with the visual world in an unsupervised manner to identify and recognize the objects that are in the world. The synapses in the neocortex are known to be influenced and changed in an unsupervised manner by the patterns of activity in sensory neurons. The unsupervised learning typically uses the observed input patterns for training and clustering the data into common features or groups. Some of the unsupervised algorithms include Hebb learning (Hebb, 1949), Boltzmann machine, and maximum likelihood (ML) density estimation method. Note: more references to unsupervised learning can be found in (Dayan, 1999; Ng, 2003).

Supervised learning is important for task-oriented type of learning, e.g. learning to drive a car or playing a game. In most cases, supervised and unsupervised learning complement one another. For example, in the model of the visual streams, most use the unsupervised learning at the lower visual cortex (such as in V1 or V2) and the supervised learning at the higher visual cortex (such as in IT).

Transfer learning. The process of applying existing knowledge into new settings is known as transfer learning. Transfer learning is enhanced by similarities between the old and new contexts.

Match and mismatch learning. Professor Stephen Grossberg believes that learning within sensory and cognitive domain is often match learning.[a] Match learning occurs only if a good enough match occurs between bottom-up information and a learned top-down expectation that is read-out by an active recognition category, or code. When such an approximate match occurs, previously learned knowledge can be refined. If novel information cannot form a good enough match with the expectations that are read-out by previously learned recognition categories, then a memory search, or hypothesis testing, is triggered, which leads to the selection and learning of a new recognition category rather than the catastrophic forgetting of an old one.

In contrast, learning within spatial and motor processes is proposed to be mismatch learning that continuously updates sensory-motor maps or the grains of sensory-motor commands. As a result, we can stably learn what is happening in a changing world, thereby solving the stability-plasticity dilemma, while adaptively updating

[a] Professor Stephen Grossberg called it "match-based learning."

our representations of where objects are and how to act upon them using bodies whose parameters change continuously through time.

Language learning. Language is probably the most complex skill a human has to learn. The ability to speak involves the fine control of many muscles. It is believed that this ability is somewhat pre-wired to some extent. An infant will start to learn to imitate another human's speech to learn the language.

Summary

In summary, learning allows us to adapt our behaviors to the environment. At the cellular level, learning occurs due to neural plasticity and involves interactions among the motor, sensory, and memory systems. The brain, with its plasticity and stability, is a great wonderful living machinery that enables the various human learning discussed in this chapter. Although there are still many research gaps with regards to the biological learning mechanism, there are many good principles we can learn from the biological human learning. These include:

- *Attention process.* We do not learn everything, i.e. our brain pays attention and learns only the important things.
- *Passive learning.* Our brains also have an automated process of passive learning due to sensory exposure, which can directly speed up our learning process on essential events.
- *Forgetting process.* We forget, and that in some sense frees up or does not unnecessary burden us with large amounts of useless or non-critical data that do not affect our life very much. It is observed that the human brain will forget things much faster if no attention is given. Note that this is besides the point highighted on passive learning discussed in perceptual learning for visual stimulus.
- *Automatic process.* Processes once learned are often automated. For example, after we learn to walk or eat, we do not need to consciously control every moment, for it becomes an automatic process. Another example is cycling: after you have learned to

cycle, the motor action has become automatic; this leaves you to focus on interesting and/or unexpected events that may happen as you cycle. This automatic process frees our brain to process other information.

Lastly, I conclude with Minsky's statement of learning from mistakes. "*I'll bet that when we try to make machines more sensible, we'll find that learning what is wrong turns out to be as important as learning what's correct. In order to succeed, it helps to know the likely ways to fail. Freud talked about censors in our minds; that keeps us from forbidden acts or thoughts. And, though those censors were proposed to regulate our social activity, I think we use such censors, too, for ordinary problem solving — to know what not to do. Perhaps we learn a new one each time anything goes wrong, by constructing a process to recognize similar circumstances, in some subconscious memory.*" — Marvin Minsky 1982, "Why people think computer cannot."

Chapter 6

EMOTION AND COGNITION

Introduction

Why is emotion related to cognition? Nobel laureate Herbert Simon emphasized that a general theory of thinking and problem solving must incorporate the influences of emotion (Simon, 1967). Psychological research has revealed that there is a direct behavioral link between cognition and emotion, i.e. our thoughts affect the way we feel about stimuli and vice versa. Emotion theorists have argued that emotions play an important role in cognition, influencing perception and creativity, and also act as a powerful motivator. Neuroscientists and biologists have also observed the link between the amygdala (center of emotion chemical secretion), the limbic system (which encompasses the amygdala and is involved in emotion and motivation), and the neocortex (center of higher intelligence). Some scientists have argued that the demands of a system with finite resources operating in a complex and unpredictable environment naturally give rise to the need for emotions, to address multiple concerns in a flexible, intelligent, and efficient way.

However, there are also concerns that emotion is not logical and strong emotion may impair rational decision-making. Acting "emotionally" implies acting irrationally, with poor judgment. Professor Picard, MIT media lab, explains that emotion not only contributes to a richer quality of interaction, but also directly impacts a person's ability to interact in an intelligent way (Picard, 1997, Affective Computing). For example:

- Communication. Emotional skills are important to communication; this includes the way we express ourselves in body action, facial expression, verbal tone, and choice of words.
- Fast decisions. Emotions help us to make fast decisions in crucial situations. For example, when one sees a tiger approaching, the fear will force one to make a decision to run or do something quickly to avoid falling prey to the tiger. Through the release of endocrines, i.e. adrenaline, which leads to faster heart beats and enables the person to respond to the situation with sharp focus. Hence, in these kinds of situations, emotions help induce fast decision-making and act like a protective mechanism.

Therefore, emotions are not as irrational as they seem to be. Picard indicates that emotion plays an active part in intelligence, especially perception, rational thinking, decision-making, planning, creativity, and so on. Studies have shown that damage to the amygdala is associated with impairment in decision-making (Bechara *et al.*, 1999) and other complex cognitive and behavioral functions (Bechara *et al.*, 2003; Bar-On *et al.*, 2003). Collectively, this demonstrates that emotion and cognition are closely coupled and suggest that emotion has a strong, pervasive, and controlling influence over cognition.

> "*Emotions arise from the relationship between the individual and its environment, or, better, the regularities of its environment. Cognition is a pre-requisite for emotion.*" — Richard Lazarus (University of California, Berkeley).

In this chapter we will discuss:

- What are the emotions?
- What is the possible emotional circuitry in the brain?

- What is emotional state?
- How can we model emotion and cognition?
- How do emotions affect our cognitive abilities?

One of the key references in this chapter is Professor Rosalind Picard's book entitled, *Affective Computing*, published in 1997. I am fortunate to have sat in her course on Autistic and Technology for fall 2007 in MIT Media Lab. Other sources of information in this chapter arise from discussions with Dr Maurizio, a researcher from MIT Brain and Cognitive Science Department.

What are the Emotions?

There are three possible categories of emotions, namely (Damasio, 1994; LaBar and LeDoux, 1996):

- Primary emotions
- Secondary emotions
- Background emotions

Primary emotions are our first response to a situation. For example, if we are threatened, we may feel fearful. These are instinctive responses without going through the thinking process. It is the more reactive part. Primary emotions can include shame, guilt, fear, anger, sadness, happiness, and embarrassment, as well as feeling insulted, pressured, and cheated.

Secondary emotions appear after primary emotions. These involve more thought and interpretation. Secondary emotions may be caused directly by primary emotions. For example, the fear of a threat may turn to anger after going through the thinking process and one realizes it is an unjustified threat. Secondary emotion gives you a picture of the person's mental processing of the primary emotion.

Background emotion is the emotional resting state or homeostasis (first suggested by Damasio). It is what you experience most often, what you feel in-between bursts of happiness and anger, interest and despondency. Background emotion is like a small power generator monitoring the condition of your body and brain as you go about your business. When the generator sings out or groans, reflecting a change in your visceral and/or musculoskeletal state, you become immediately aware of the change and experience a primary or secondary emotion.

What is the Possible Emotional Circuitry in the Brain?

History of the emotional circuit. In 1937, James Papez first conceived the mechanism for emotion. The Papez circuit consists of the neural pathway that goes from the hypothalamus, thalamus to hippocampus, and spreads to the cingulated cortex, which provides the link to the neocortex. Papex thought this the cortical control of emotion. See figure below (Figs. 6.1 and 6.2).

Papex's proposed circuit (Fig. 6.1), however, missed out the amygdala, which we now know is the center of the emotion.

In 1949, Paul McLean extended Papez's circuit to include the amygdala and more explicit connections between the hypothalamus,

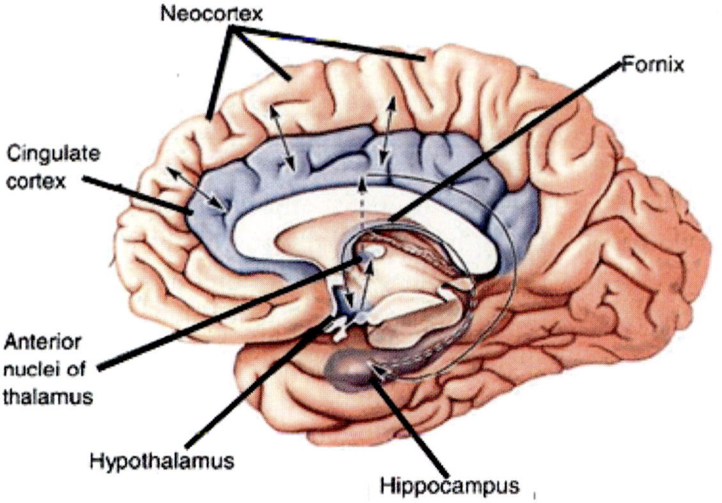

Fig. 6.1. Papez's proposed circuit.

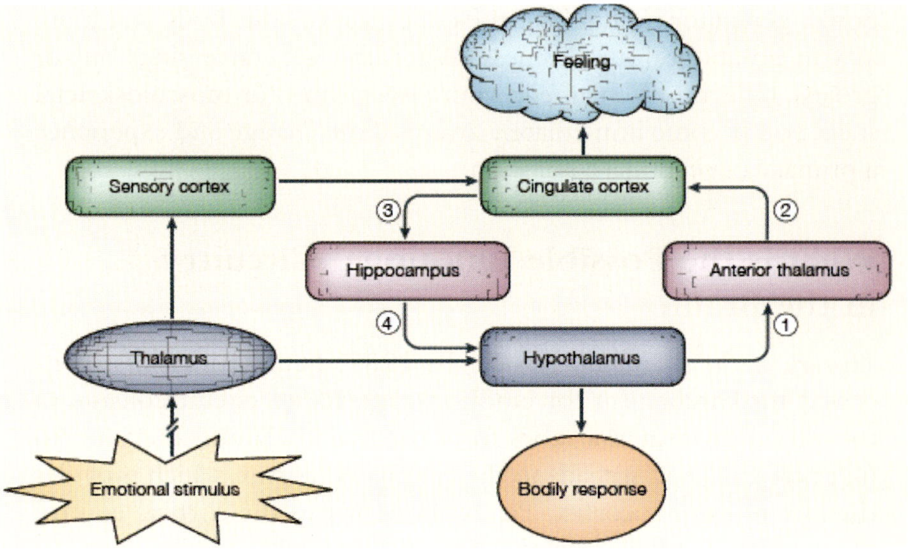

Fig. 6.2. Papez circuit in schematic diagram. (Picture from Dalgleish, 2004.)

hippocampus, and the links from the prefrontal area to the anterior area of the brain and the brainstem.

Figure 6.3 shows Paul MacLean extension of the Papez's circuit in 1949 to include the amygdala and other areas that feed from the Papez regions. The amygdala is the key player in emotion as we know today.

Figure 6.4 shows the structure that is often considered to constitute the limbic system. The limbic system was introduced as a concept by Paul MacLean in 1952 and was long considered the seat of emotions. Though some of the structures included in this system are in fact involved in some emotional responses, we now know that it does not correspond exactly with any of the multiple emotional systems in the brain. Picture and description adapted from http://thebrain.mcgill.ca/.

Amygdala. The amygdala is located deep within the medial temporal lobes of the human brain. Research shows that the amygdala performs a primary role in the processing and memory of emotional reactions. The amygdala is part of the limbic system. It

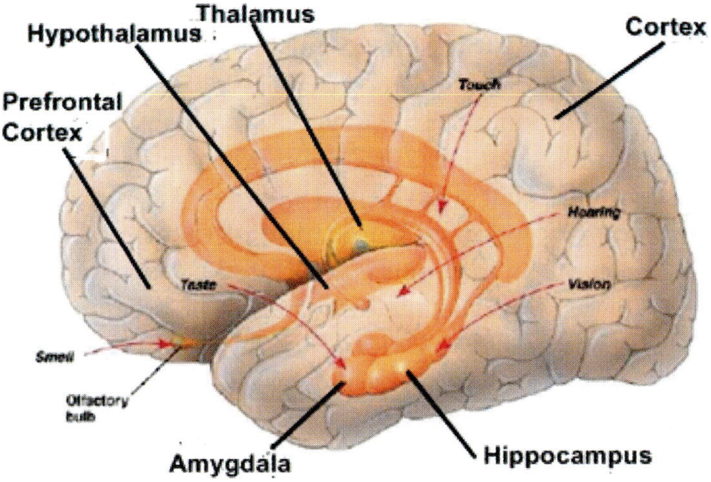

Fig. 6.3. Paul MacLean's proposed circuit.

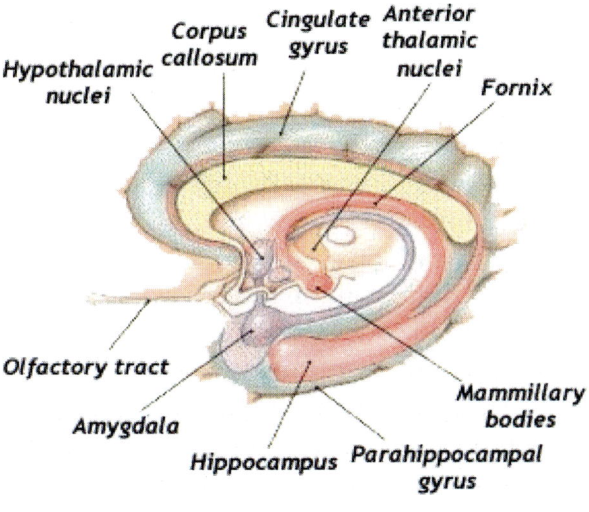

Fig. 6.4. Details of the limbic system. (Picture adapted from http://thebrain.mcgill.ca/.)

connects with the hippocampus, the prefrontal area, and the medial dorsal nucleus of the thalamus. What would happen if a person loses the function of the amygdala? A patient named "S.M." had a Urbach-Wiethe disease, which causes bilateral calcification of the amygdala. Such patients do not experience anger or fear (are unable to recognize expressions), and have diminished emotional responses. Later studies confirmed that amygdala lesions in humans abolish fear conditioning. The amygdala is involved in processing emotional stimuli (Armony, 2007), and it is where the learning for fear conditioning occurs (LeDoux, 1994).

Thalamus. The lesion or stimulation of the medial dorsal and anterior nuclei of the thalamus is associated with changes in emotional reactivity. However, the importance of these nuclei on the regulation of emotional behavior is not due to the thalamus itself, but to the connections of these nuclei with other limbic system structures. The medial dorsal nucleus makes connections with cortical zones of the pre-frontal area and with the hypothalamus. The anterior nuclei connect with the mamillary bodies, and through them, via the fornix, with the hippocampus and the cingulate gyrus, thus taking part in the Papez's circuit.

Hypothalamus. Lesions of the hypothalamic nuclei interfere with several vegetative functions and some of the so-called motivated behaviors, like thermal regulation, sexuality, combativeness, hunger, and thirst. The hypothalamus is also believed to play a role in emotion. Specifically, its lateral parts seem to be involved with pleasure and rage, while the median part is likely to be involved with aversion, displeasure, and a tendency for uncontrollable and loud laughter. However, in general terms, the hypothalamus has more to do with the expression (symptomatic manifestations) of emotions than with the genesis of the affective states. When the physical symptoms of emotion appear, the threat they pose returns, via the hypothalamus, to the limbic centers, and thence, to the pre-frontal nuclei, increasing anxiety. This negative feedback mechanism can be so strong as to generate a situation of panic. As will be seen later on, the knowledge of this phenomenon is very important, for clinical and therapeutic reasons.

Cingulate gyrus. It is located in the medial side of the brain between the cingulate sulcus and the corpus callosum (principal fiber bundle connecting the two cerebral hemispheres). There is still much to be learned about this gyrus, but it is already known that its frontal part coordinates smells and sights with pleasant memories of previous emotions. This region also participates in the emotional reaction to pain and in the regulation of aggressive behavior. Wild animals, submitted to the ablation of the cingulate gyrus (cingulectomy), become totally tamed. The cutting of a single bundle of this gyrus (cingulotomy) reduces pre-existent depression and anxiety levels, by interrupting neural communication across the Papez's circuit.

Brainstem. The brainstem is the lower part of the brain. It is believed that the brainstem is responsible for the "emotional reactions", (indeed, they are just reflex answers) of inferior vertebrates, like reptiles and amphibians. The involved structures are the reticular formation and the locus coeruleus, a concentrated mass of nor-epinephrine secreting neurons. It is important to stress that, even in humans, these primitive structures remain active, not only as alerting mechanisms vital for survival, but in the maintenance of the sleep-awake cycle.

Ventral Tegmental Area (VTA). The VTA is part of the midbrain (see Fig. 6.5), located in the mesencephalic part of the brainstem. In VTA, there is a compact group of dopamine-secreting neurons whose axons end in the nucleus accumbens (mesolimbic

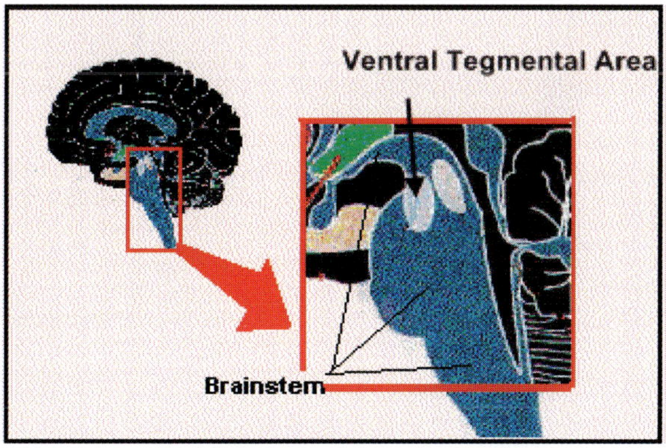

Fig. 6.5. The brain stem area. (Picture adapted from http://www.cerebromente.org.br/.)

dopaminergic pathway). The spontaneous firing or the electrical stimulation of neurons belonging to that region produces pleasurable sensations, some of them similar to orgasm. Many people who, for a genetic error, have a reduction of D2 (dopamine) receptors in the accumbens nucleus, become, sooner or later, incapable of obtaining gratification from the common pleasures of life. Thus, they seek typical and noxious "pleasurable" alternatives, like alcoholism, cocaine addiction, impulsive gambling, and compulsion for sweet foods. Information from http://www.cerebromente.org.br/.

Septum (or septal nuclei). The septum part of the limbic system lies anteriorly to the thalamus (Fig. 6.6). The septum area has been associated with different kinds of pleasant sensations, mainly those related to sexual experiences.

Prefrontal cortex. When the pre-frontal cortex suffers a lesion, the subject loses his sense of social responsibility as well as the capacity for concentration and abstraction. In some cases, although consciousness and some cognitive functions, like speech,

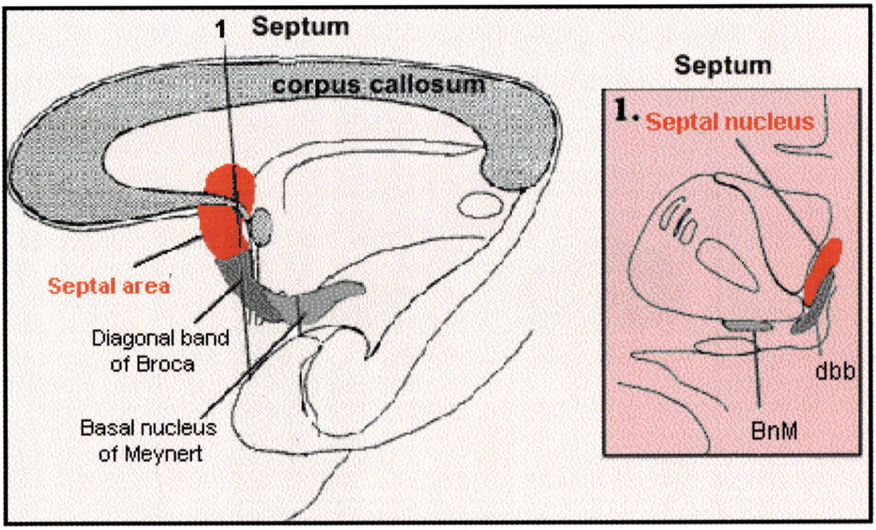

Fig. 6.6. The septum area. (Picture adapted from http://www.cerebromente.org.br/.)

remain intact, the subject can no longer solve problems, even the most elementary ones. When pre-frontal lobotomy was used for treatment of certain psychiatric disturbances, the patients entered into a stage of "affective buffer," no longer showing any sign of joy, sadness, hope, or despair. In their words or attitudes, no traces of affection could be detected.

A critical finding was that information reaches the amygdala before the cortex. In fact, some stimuli may never reach the cortex at all. That is why we may react to a situation without even realizing what we are doing until well after we have done it.

Sometimes, the brain lesions that invoke particular emotional changes are also responsible for cognitive impairments. This condition is clearly illustrated in patients of the Klüver–Bucy Syndrome. This disorder results from bilateral lesions of the amygdala and inferior temporal cortex. One emotional symptom associated with the Klüver–Bucy Syndrome is a "flat" affect, i.e. the patient becomes indifferent to people or situations, showing very little affective behavior to stimuli that normally arouse emotions. Paradoxically, the patient may occasionally smile inappropriately. There are cognitive abnormalities that accompany these emotional symptoms. For example, the patient suffers from visual agnosia and is unable to recognize faces. The presence of emotional and cognitive symptoms together following damage to the amygdala and inferior temporal cortex may imply that these brain structures are part of the neural networks for both cognitive and emotional processes (Kolb and Whishaw, 1996).

The sensory input to the prefrontal cortex is highly integrated with input from cortices associated with emotional processes. The following diagram illustrates the connection.

Figure 6.7 shows the direct and possible indirect pathways of sensory projections to lateral and orbital prefrontal cortices. Lateral prefrontal areas, such as the frontal eye fields, receive robust projections from visual cortices and sparse projections from the visual-recipient part of the amygdale. In contrast, caudal orbitofrontal areas receive projections from olfactory, gustatory, auditory,

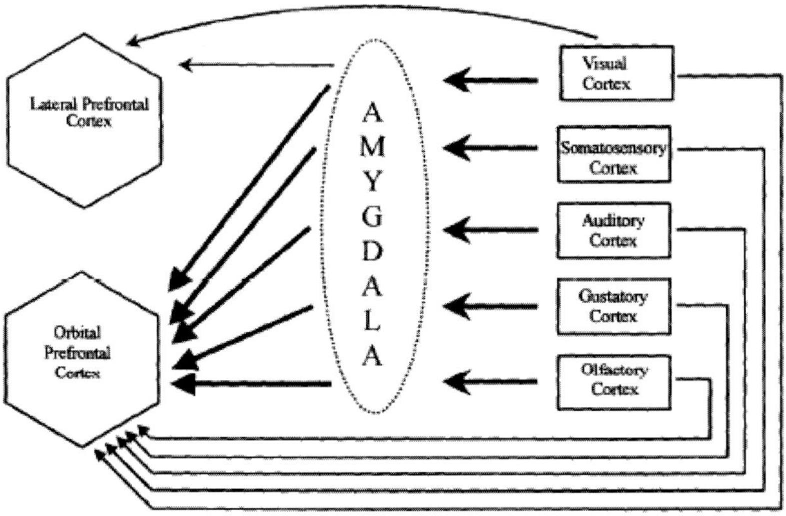

Fig. 6.7. Direct and indirect pathways of sensory projections to prefrontal cortices.

somatosensory, and visual cortices, and robust projections from the amygdale, which is also the recipient of input from the same sensory modalities. Information from (Barbas, 1995).

Figure 6.8 shows the information from an external stimulus reaches the amygdale in two different ways. One is a short, fast, but imprecise route directly from the thalamus; and another is a long, slow, but precise route, by way of the cortex. It is said that the short, more direct route lets us start preparing for a potential danger before we even know exactly what it is. For example, let's suppose while you are walking in a forest you suddenly see a long, narrow shape coiled at your feet. This snake-like shape, very quickly, via the short route, sets in motion the physiological reactions of fear that are so useful for mobilizing you to face danger. But this same visual stimulus, after passing through the thalamus, will also be relayed to your cortex. A few fractions of a second later, the cortex, thinking with its discriminatory faculty, will realize that the shape you thought was a snake was really just a discarded piece of garden hose. Information from http://thebrain.mcgill.ca/.

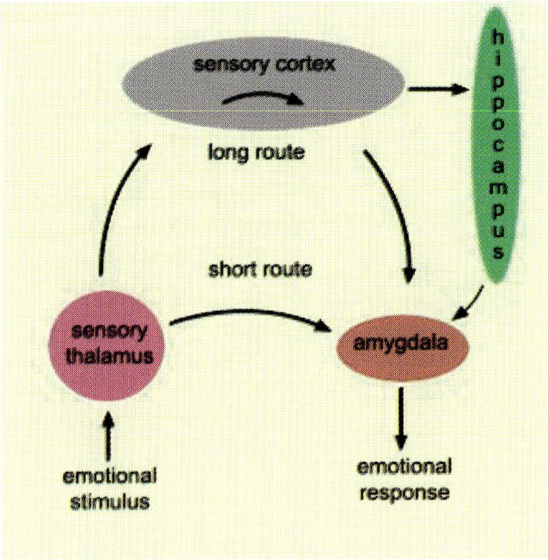

Fig. 6.8. External stimulus reaches amygdale in two different routes. (Picture from http://thebrain.mcgill.ca/.)

Figure 6.9 shows a clear link between the prefrontal cortex (location well known for working memory), neocortex (the central for higher intelligence, i.e. cognitive abilities), hippocampus (responsible for long term memory), and amygdala (responsible for emotional chemical secretion). Information from (LaBar and LeDoux, 2006).

Remark: We can see that emotion is not a separate subsystem of the brain (or mind *per se*) but a pervasive feature of it. It is important to stress that all these structures interconnect intensively and none of them is solely responsible for any specific emotional state. However, some contribute more than others to certain kinds of emotion.

The process of emotion engages parts of the cortex, in particular the frontal cortex. The frontal cortex "communicates" significantly with the limbic system. Damage to this area impairs normal cortical-limbic interaction, effectively leaving a person with too little emotion. And too little emotion impairs decision-making.

We know that too much emotion can wreak havoc on reasoning, but now there is evidence that too little emotion also can wreak havoc. This evidence requires a shift from the usual notion of how people separate emotions and rationality. Detailed studies

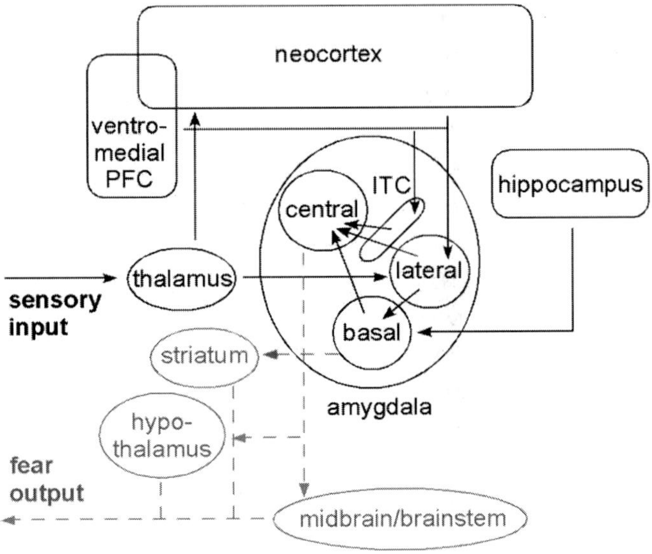

Fig. 6.9. A proposed circuitry for the fear emotion.

have been done by Damasio on patients who have frontal-lobe disorders, affecting a key part of the cortex that communicates with the limbic system. For example, a healthy individual who loses a lot of money with an investment would learn the lesson that investment is not a simple business and might stop investing; however patients with too little emotion might continue to invest until all their money is gone. Another example is a patient who suffers from frontal lobe damage, and who may go into an endless irrational search when engaged with a simple decision such as when to schedule an appointment. Another effect frontal-lobe disorders have on patients is that they might not have the usual feelings of embarrassment or bad feeling (positive or negative feeling with regards to certain decisions.) Feelings are always part of decision-making.

What is Emotional State?

Emotional state refers to your internal dynamics when you have an emotion. The state is multi-variant, including aspects of both your mental state and physical state. It changes with time and with a variety of other activating and conditioning factors. Emotional state cannot be directly observed by another person, but may be inferred (Picard, 1997). For example, when you are standing in line, you do not know the emotional state of the person in front of you. However, when he turns around with a clenched fist, you can infer that he is in a state of anger. When we change our emotional states, we are switching between different ways to think (Picard, 2004). For example, emotional states influence:

- What information is available in working memory (Bower, 1991)
- Subjective utility of alternative choices (Lerner and Keltner, 2000)
- Style of processing (Bless *et al.*, 1996)

Primary emotions are our first emotion state. They are our first response to a situation.

Physical aspects of different emotion states. Typical emotional states are expressed in physical forms, such as posture, vocal intonation, facial expression, and limb movement. Facial expression is the most noticed area of emotional expression.

The face is where our eyes linger during conversation. We tend to communicate most effectively "face to face". Facial expression provides us many emotional cues of the person you are talking to. Professor Paul Ekman, University of California, termed "social display rules" that limit the range of acceptable expressions, such as in business or social settings. For example, during a serious negotiation session, it is inappropriate for the businessman to distort his face to extreme disgust. He knows that in a serious meeting, he can express only mild emotions regardless of his feeling.

Autistic children who have social problems avoid facial contact. They have difficulty in generalizing facial expressions and understanding why or what causes that facial expression. And hence, this affects some autistic children's cognitive processes with regards to emotional intelligence. Computers today are somewhat like autistic people. They are factual and do not understand the emotional aspect of the situation.

Emotion affects the body's effectiveness in function. Studies have long shown that emotions such as anxiety, fear, and stress influence your body; in particular the likelihood that you get sick increases as emotions affect the immune system. Research has found mechanisms whereby emotion directly influences the immune system, neurochemically, as well as through regulation of the autonomous nervous system, which have been found to directly interact with cells of the immune system such as lymphocytes and macrophages. Hormones that are released during stress have also been found to impact immune cells.

An experiment has shown that babies use their emotional states to make intelligent decisions. For example, babies are reluctant to crawl on a glass surface if the apparent drop beneath them is large (this seems to be an innate safety mechanism to keep babies away from dangerous heights). However, if the drop is a little smaller, so that the situation is ambiguous, the babies use social referencing to judge whether or not to cross. In the experiment, the parent sits opposite the baby, and is asked to adopt either a happy or fearful expression. Surely enough, the baby will cross when the parent is happy but not when the parent is fearful. This is interesting because it is the opposite of the typical response in normal interaction, where a fearful expression will normally cause a child to approach the parent for security. Clearly, these babies were using emotional information from their parents to make sense of the ambiguous situation.

> "*Our earliest emotions are built-in processes in which inborn proto-specialists control what happens in our brains. Soon we learn to overrule those schemes, as our surroundings teach us what we ought to feel. Parents, teachers, friends and finally our self-ideals impose upon us new rules for how to use the remnants of those early states: they teach us how and when to feel and show each kind of emotion sign. By the time we are adults, these systems have become too complicated to understand. By the time we have passed through all those stages of development, our grown up minds have been rebuilt too many times to remember or understand much of how it felt to be an infant.*" — From Society of Mind, Malvin Minsky.

Figure 6.10 shows Plutchik Robert's three-dimensional circumplex model that describes the relations among emotion concepts, which are analogous to the colors on a color wheel. The cone's vertical dimension represents intensity, and the circle represents degrees of similarity among the emotions. The eight sectors are designed to indicate that there are eight primary emotion dimensions (Fig. 6.11) defined by the theory arranged as four pairs of opposites. In the exploded model, the emotions in the blank spaces are the primary dyads — emotions that are mixtures of two of the primary emotions (Plutchik, 2002).

Another perspective of looking at human emotion is Russell's model (Russell, 1980; Russell and Barrett, 1999).

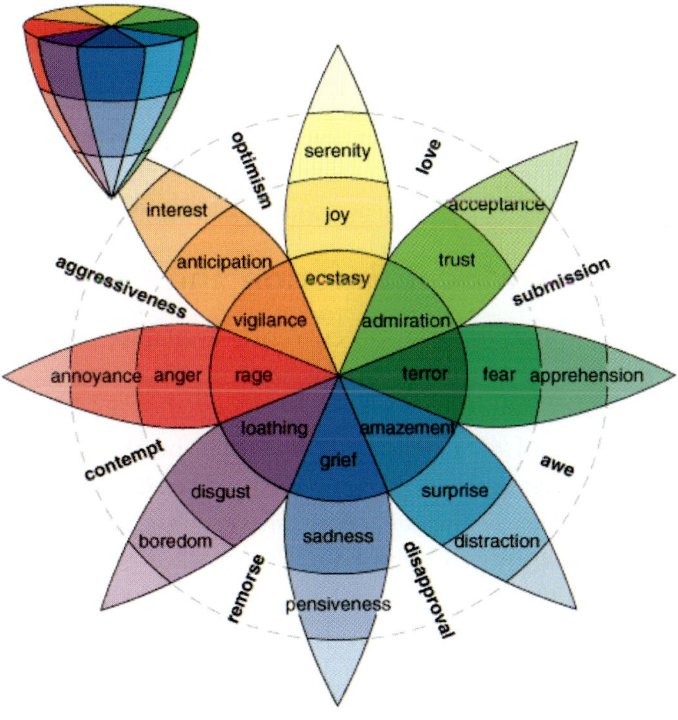

Fig. 6.10. Plutchik Robert's three-dimensional circumplex model.

stimulus event	cognition	feeling state	overt behavior	effect
threat	"danger"	fear	escape	safety
obstacle	"enemy"	anger	attack	destroy obstacle
gain of valued object	"possess"	joy	retain or repeat	gain resources
loss of valued object	"abandonment"	sadness	cry	reattach to lost object
member of one's group	"friend"	acceptance	groom	mutual support
unpalatable object	"poison"	disgust	vomit	eject poison
new territory	"examine"	expectation	map	knowledge of territory
unexpected event	"what is it?"	surprise	stop	gain time to orient

Fig. 6.11. The general link between the eight primary emotions in humans, considered by Plutchik, namely, anger, fear, sadness, disgust, surprise, curiosity, acceptance, and joy.

How Can We Model Emotion and Cognition?

What are the models of emotion currently available? Four different types of models are discussed in this section. These are:

The Ortony Clore Collins (OCC) Cognitive model. In 1988 Ortony, Clore and Collins published their book entitled, "The Cognitive Structure of Emotions" (Ortony *et al.*, 1988). They set forth one of the early models of cognitive appraisal for emotions that has come to known as the OCC model. The OCC model's main goal is for the AI system to be able to reason about emotions especially for natural language understanding, cooperative decision solving, and planning. The original OCC model has been extended by others to generate emotion (although the original OCC structure was formulated without having emotion generation in mind).

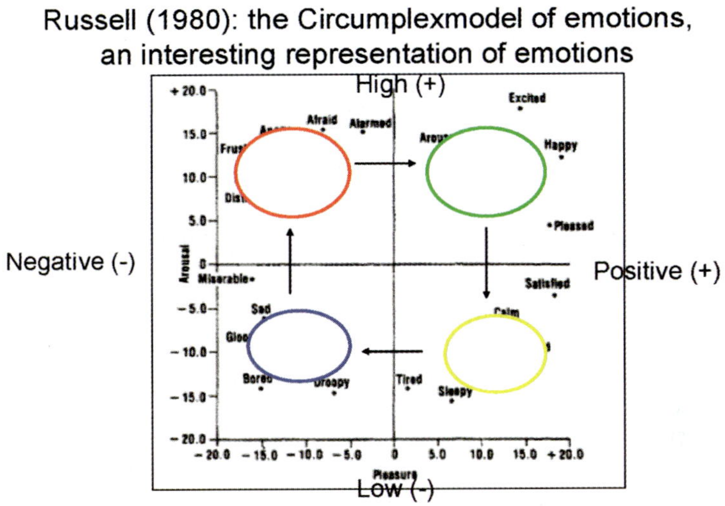

Fig. 6.12. Russell's two-dimensional representation of emotions.

By cognitive reasoning, the computer can deduce that a sequence of events causes an emotion to arise. By the same reasoning, applied to its personal events, it can cause an emotion to arise within itself. Hence, the OCC structure can be used not just for reasoning about emotions, but also for generating them.

Roseman's Cognitive Appraisal Model. Roseman and colleagues postulated that cognitive appraisals of events are based on six specific dimensions, namely (McCarthy *et al.*, 1998):

- Situational state — an appraisal of whether an event is consistent or inconsistent with one's desires.
- Motivational state — which refers to whether the individual is seeking something positive or striving to avoid something painful.
- Probability — which refers to the perceived likelihood of an event's occurrence.
- Control potential — the degree to which individuals believe they can control or influence a given situation.
- Problem source — which refers to whether a negative event is caused by something inherent to the person or object or merely

with the behavior or a non-central attribute of the person or object.

- Agency — which consists of three separate sub-dimensions
    - Agency-self: the degree to which an event is perceived as caused by oneself.
    - Agency-other, the degree to which the event is perceived as caused by another person.
    - Agency-circumstance, the degree to which the event is perceived as caused by external circumstances.

For example, John aims to earn an "A" grade, but it is uncertain if he will do well enough on the final exam to receive one. His motivational state is appetitive, aiming for a reward. His situational state is presently uncertain. The causal agency is a test (impersonal). He has been working hard and thinks he has potential to receive an "A". His appraisal of his situation suggests that he feels hopeful. If he then receives his grade and it is not an "A" (motive-inconsistent), then he may feel frustration. If he feels that his failure on the test was due to the teacher's grading him unfairly, then he is likely to feel anger towards his teacher.

The model suggests that appraisals are influenced by shifts in attention. If John wants an "A" and gets one, he may focus on the "A" and feel joyful. Or, he may think about the teacher and feel a liking for him, or he may focus on what he accomplished herself and feel pride. Picard pointed out that teaching a computer what to attend to is another open research problem. In humans, attention is influenced by emotion. For example, anger can focus attention on the object of the anger.

Roseman and colleagues found that by measuring appraisals along each of these dimensions, an individual's emotional reaction can be predicted. The theory includes 11 specific negative emotions and five positive emotions. The 11 negative emotions are disgust, distress, sadness, fear, unfriendliness, anger, frustration, shame, regret, and guilt. The five positive emotions are joy, relief, affection, pride, and hope.

Both the OCC and Roseman theories provide a rule-based mechanism for cognitive generation of emotions. And they address the emotional states at a relatively high-level. In humans, emotions are generated not only by explicit reasoning (high level), but also by low-level non-cognitive influences, such as physical influences, i.e. associated with bodily phenomena rather than mental phenomena.

Three-layer architecture. Professor Aaron Sloman (University of Birmingham) and colleagues have proposed that adult humans have at least three architectural layers in their brains:

- A reactive layer — executes in almost an automatic manner. The reactive layer is capable of some simple learning; however, it is not able to construct or evaluate plans. (suitable for fast primary emotions)
- A deliberative layer — capable of planning, evaluating options, making decisions, and allocating resources. The emotions involved in goal-success or goal-failure. (This corresponds to Damasio's secondary emotions.) These are cognitively-generated emotions that typically require some kind of cortical reasoning about goals, situations, objects, and events.
- A self-monitoring layer — is a self-monitoring meta-management that prevents certain goals from interfering with each other and can look for more efficient ways for the deliberative layer to operate, choose strategies, and allocate its resources. Sloman suggested that emotions associated with this layer might include shame, humiliation, and grief.

Although the three-layer architecture lacks details of implementation, it illustrates the need Picard argued for multiple levels of models in emotion synthesis, including both low-level primary mechanisms and higher-level cognitive ones. In particular, it illustrates the need for a higher "self-monitoring" process for the management of emotions, to develop the skills of emotional intelligence for regulating and wisely using its emotions.

Picard indicated that the three-layer architecture is a potential model for emotion synthesis that compares favorably with findings in the neurological, psychological, and cognitive science communities.

These three layers can be categorized loosely according to their functional similarity with other animals. See also the six-layer architecture proposed by Marvin Minsky (Minsky, 2006) (discussed in Chapter 9).

Cathexis — four elicitors for emotion synthesis. Carroll Izard (1993) proposed that there are four types of elicitors of emotion in humans. These have inspired a new connectionist model of emotion synthesis. Juan Velasque of MIT (Velasquez, 1996) has used these four types of elicitors to build an emotional synthesis called "Cathexis". The four elicitors in this model are:

- Neural — effect of neurotransmitters and other neurochemical processes. These processes run independently, in the background, and are influenced by hormones, sleep, diet, depression mediation, etc.
- Sensorimotor — effect of posture, facial expression, muscular tension, and other central efferent activity. These effects primarily intensify a given emotional state, but in some cases appear capable of generating new affective states.
- Motivational — effect of sensory provocations such as anger provoked by pain of drives such as hunger and emotions evoking each other.
- Cognitive — effect of cortical reasoning, implemented here via an adaptation of Riesman's theory.

Cargexis consists of a constellation of proto-specialists, like Minsky's agents in the Society of Mind (Minsky, 1985). Each proto-specialist represents a basic emotion types, which receives inputs from the four elicitors, as well as from other proto-specialists. Each proto-specialist can exert influence on output behaviors, e.g. joy with an intensity above its activation threshold, can produce a smile or joy, and can inhibit distress and activate hope. Since proto-specialists are used to implement both emotional and non-emotional states, it is easy for emotions to interact with physical states; for example, sorrow increases fatigue and decreases hunger. The result is a distributed connectionist-flavor model that can synthesize a variety of emotions simultaneously.

The Cathexis model has only one update rule. The rule contains terms that take on values specific to proto-specialists, but otherwise the form is the same for every proto-specialist's emotion intensity (see annex for detail updating rules).

Interested readers on the four models can refer to the references for more information.

Learning from autistic children may give us insights to building a model for emotions. Autistic children do not interpret socio-affective cues such as tone of voice or facial expression well. Their social inability to perceive, understand, and act using social-emotional information affect their ability to interact and communicate effectively with people. Today's computer is naturally an autistic system in the emotional aspect. Hence, how can a system be built to be less autistic? Learning from autistic children and learning how to help them to improve their social-emotional skill may in turn help us to understand such social-emotional cues and make a system that better responds to humans and learns like a human.

> "*...process of reward and punishment expresses itself through emotion. Emotions are "states produced by reinforcing stimuli". "Reinforcing" stimuli are the ones that need to be decoded by special regions of the brain, which generate the neural processes that we call "emotions". The representation of a "reinforcer" contains information about how to calculate reward/punishment. The brain wants to maximize the activation of representations relative to rewarders and minimize activation of the representations relative to punishers.*" — Edmund Rolls (University of Oxford).

How Do Emotions Affect Our Cognitive Abilities?

Below are four examples of how emotion can affect our cognitive abilities.

1. *Emotion in decision-making.* Picard suggested that affective decision-making provides a good solution in the situation where problems are faced, in which the possibilities cannot be enumerated and evaluated in the available time. Picard added that humans use feelings to help them navigate the oceans of inquiry, to make decisions in the face of combinatorial complexity. These feelings might be called "intuition" or "a sense of knowing" or just "gut feelings". Regardless of what they are called, they provide a mechanism through which emotion works powerful influences on human cognition and behavior. It is commonly observed that humans can respond with remarkable intelligence and flexibility despite insufficient knowledge, limited memory, and relatively slow processing speed. An integral component of human decision-making is emotion, and this component could potentially be given to computers.
2. *Emotions in learning.* Learning can be abandoned because of negative feelings or enhanced because of positive feelings. Dr Barry Kort suggested that learning systems have affective states, and that future learning systems should exhibit such learning states as curiosity, fascination, puzzlement, frustration, insight, satisfaction, and confidence.
3. *Emotions influence memory.* Emotions influence our memory retrieval and memory encoding. Good feelings likely encode a knowledge of positive outcomes. Bad feelings likely to encode a knowledge of bad outcomes. Likewise, positive moods make it easier to remember positive things while negative moods tend to make it easier to remember negative things. For example, when you come across a car accident, you may remember your own negative experience of a car accident. Also, we tend to

remember emotional events better, regardless of whether the emotion was good or bad. For example, experiments have shown that faces with fearful expressions are remembered better than those with neutral ones.

Although memory retrieval remains a mystery, we know that emotion plays a role in its function. Picard in his book, "Affective computing", indicated that memory may be the chief mechanism through which emotions enter into the mental associations active in analogical thinking and creativity. Picard saw the influence of emotions in both high-focus reasoning and low-focus generation of associations as a consequence of emotional influence on memory.

4. *Negative emotion can impair our cognition.* Too much prolonged negative emotion does not seem to help in cognition. For example, studies have shown that depression can raise the risks of developing dementia and Alzheimer's disease later in life. Dementia is the progressive decline in cognitive function of the brain. The areas affected include memory, attention, language, and problem solving. Alzheimer's disease (AD) is a neurodegenerative disease characterized by progressive cognitive deterioration. Some scientists believe that depression releases harmful stress hormones, which may damage brain structures responsible for memory such as the hippocampus. (Source: Archives of general psychiatry (data)).

Conclusions

The brain has specific circuits that handle emotions. These circuits help to communicate "emotion" to the rest of the body via the bloodstream and the nervous system. Emotion is a mechanism that regulates the body. It causes changes in the body state. Psychological research has revealed that there is direct behavioral link between cognition and emotion, i.e. our thoughts affect the way we feel about stimuli and vice versa. Our thoughts can change our emotions. This influence can be considered to be "cognitive" when they involve appraisal, comparison, categorization, inference, attribution, or judgment. Emotion could be one of the most importance factors for an intelligent machine. Picard commented that emotions may be a key reason why artificial intelligence has failed to date (Picard, 1996).

For humans to truly be able to engage with the computer and vice versa, it is important that the machine exhibit some form of emotional expression in its cognitive process and also be able to understand some level of human emotion. That would help for humans to have a desire in engaging the machine and appreciate it for everyday use. This is irrespective of whether the machines have true feelings like humans or follow a set of rules for their emotional behavior.

Today, there are no computers with emotions that influence their decision-making and other cognitive processes to the same degree that these influences are believed to occur in people.

There are models on emotion, and four of these models have been discussed. However, till today, there is no model that can completely represent the human emotional system. There are a number of issues that one needs to overcome in order to model a complete emotional system for computer. One of the key problems is how to map emotional states to behaviors. As we have seen, fear motivates a rat to find a means of escape, but it does not automatically tell it by what means to pursue. Emotions motivate and bias behaviors, but they do not completely determine them.

Chapter 7

LAMINAR COMPUTING

What, How, and Why Laminar Computing?

Laminar computing is a term coined by Professor Stephen Grossberg from Boston University for modeling biological intelligence. Grossberg has been doing research over the past 50 years on how the brain works. Some of his research covers the learning mechanism in neural networks, the mechanism of the visual cortex, how the cerebral cortex works, and how the brain perceives things, giving rise to a mind. More recently, he has been trying to understand the laminar structure in the brain to design a computational model, which is the key interest of this chapter.

The lecture series Professor Stephen Grossberg (Fig. 7.1) (Wang Professor of Cognitive and Neural Systems and Professor of Mathematics, Psychology, and Biomedical Engineering at Boston University) gave at MIT Lincoln Laboratory on neural network technology was instrumental in motivating the laboratory to initiate the national DARPA study on neural networks. More detailed articles of Grossberg's work can be found at http://www.cns.bu.edu/Profiles/Grossberg/.

This chapter discusses laminar computing, several important properties that underline laminar computing, how the ideas about laminar computing evolved from Grossberg's early research work, and how the laminar architecture research has progressed from vision to cognition.

Fig. 7.1. Picture of Professor Stephen Grossberg.

The Neocortex has a Laminar Pattern

The neocortex is typically made up of six layers of cells. In 1909, the German anatomist, Korbinian Brodmann, identified more than 50 different areas of the neocortex (Brodmann, 1909). The different areas of neocortex were identified based on the difference in thickness of these layers and the size and shape of neurons. Some examples of different areas identified in the visual pathways are the striate visual cortex (V1), prestriate visual cortex (V2, V3, V4, MT, MST), and inferotemporal cortex (IT) areas.

In humans, 90% of the cerebral cortex is neocortex, and the neocortex takes up about 76% of the human brain (Chudler, 2006). The neocortex contains about 28 billion neurons (28×10^9) according to Mountcastle (Mountcastle, 1997), and is divided into lobes. Note that if we take the average human brain as 23 billion neurons from Rabinowicz's study (Rabinowicz *et al.*, 2002), then the neocortex contains about 17.5 billion neurons (see Chapter 1). The different regions of the neocortex are believed to have different functions, such as sensory perception, generation of motor commands, reasoning, thought, language, etc.

The neocortex's six distinct layers are named as follows:

- Layer 1 — Molecular layer. This layer consists of mostly dendrites and axons of cells in deeper layers.

- Layer 2 — External granular layer. This layer comprises largely of granule cells.
- Layer 3 — External pyramidal layer. This layer has a variety of cell types, including large pyramidal neurons that project mainly to targets in other parts of the cortex.
- Layer 4 — Internal granular layer. This layer is like layer 2, composed primarily of granule cells. It typically receives inputs from layer 3 in other parts of the cortex. It can also receive direct bottom-up inputs from other cortical areas and inputs from layer 6.
- Layer 5 — Internal pyramidal layer. This layer contains mainly pyramidal cells that are typically larger than those in layer 3 and often project to subcortical targets or to layers 1/2 or 4 of lower-level parts of the cortex.
- Layer 6 — Multiform layer. This layer is a mix of different cell types and white matter.

These six cortical layers are interconnected and their connections vary with different neocortical areas, i.e. different areas in the brain perform different functions and have different layer connections. The boundaries between the layers are quite fuzzy, and it is common that neurons cross layer boundaries with their dendrites and axons. These six layers that form the neocortex represent the laminar pattern. Hence, Stephen Grossberg coined the term "laminar computing" for the "computation" performed by the laminar circuitry.

> "*Laminar computing concerns the fact that cerebral cortex, the seat of higher intelligence in all modelities, is organized into layered circuits (usually six main layers), which undergo characteristic bottom-up, top-down and horizontal interactions.*" — Stephen Grossberg (Grossberg, 2003a).

The basic structure of the cortical layers forms columns known as cortical columns in the mature cortex (some view this as a cortical maps). Hence, the neocortex is said to be organized in columnar function (Mountcastle, 1997); also known as the neocortical column.

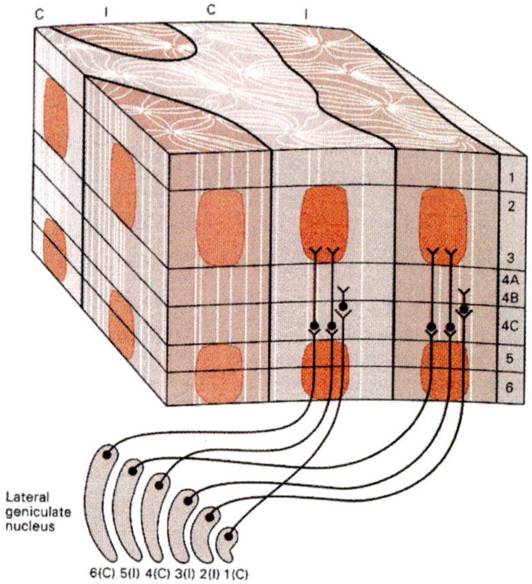

Fig. 7.2. A visual cortical column.

Figure 7.2 shows an example of a visual cortical column. Each column typically responds to a sensory stimulus representing a certain body part or region (of sound or vision). These columns are typically similar and can be thought of as the basic repeating functional units of the neocortex.

In humans, the neocortex consists of about half a million of these columns, each of which contains approximately 60000 neurons. Mountcastle also defined the cortical minicolumn as a cortical column containing 80–120 neurons. It has also been discovered that the cortical column in the visual cortex has a pinwheel-like pattern.

A pinwheel-like orientation (Fig. 7.3) has been discovered in the cortical columns of cat and monkey visual cortexes (Ohki *et al.*, 2005).

Figure 7.4 shows the arrangement of neurons, dendrites, and axons in vertical modules of the striate cortex (V1) of the macaque monkey (Mountcastle, 1997) (picture origin from (Peters and Sethares, 1996)).

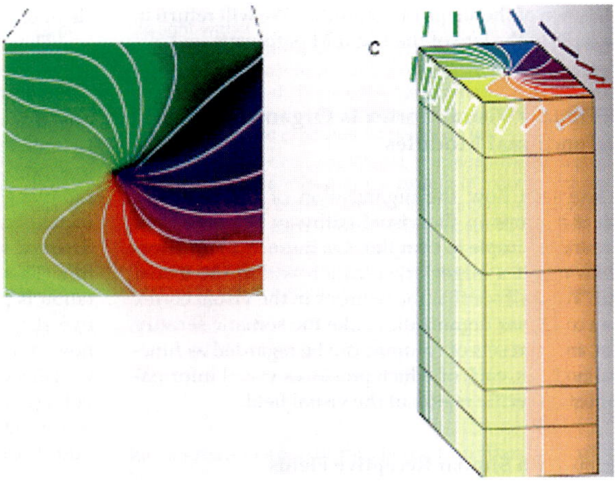

Fig. 7.3. Pinwheel representation of cortical orientation columns.

Each cortical column contains six layers of cells (Fig. 7.5). The appearance of the six distinct layers in the cortex depends on what is used to stain it. The Golgi stain reveals neuronal cell bodies and dendritic trees (extreme left picture). The Nissl method shows cell bodies and proximal dendrites (middle picture). The Weigert stain for myelinated fibers reveals the pattern of axonal distribution (extreme right picture) (Heimer, 1994).

Scientists believe that understanding the distinct functions of different parts of the neocortex, the functional utility of such laminar organization in the control of behavior should have an enormous payoff in understanding biological intelligence. Stephen Grossberg in his 2003 paper said that if one can understand how laminar circuits work in one part of the brain, then different intelligent capabilities may be expected to be understood as variation on a shared architecture theme (Grossberg, 2003a). The laminar circuits of the visual cortex are well explored by the Cognitive and Neural System Department in Boston University led by Grossberg. Research is now on-going to carry this work further from vision to cognition. As noted earlier, this chapter will discuss the laminar computing done in BU/CNS. A number of details may be missed out in the process of summarizing more than

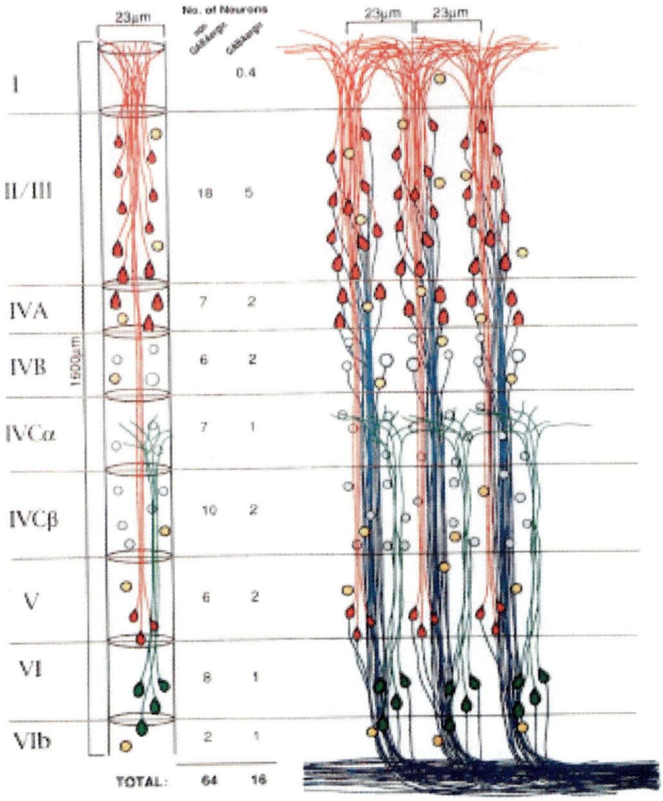

Fig. 7.4. Layer arrangement of the striate cortex (V1).

40 years of research done by Grossberg and his colleagues in this short chapter.

> "*When laminar computing models become sufficiently mature, they will provide a blueprint for VLSI chip designers to try to design a universal laminar chip set whose variations can support many different types of intelligence. These chips will, moreover, be manifestly self-consistent for integration within a larger system controller, say for an autonomous adaptive mobile robot, due to the fact that they have all been fashioned from the same underlying design, and will all connect in a stereotyped manner.*" — Stephen Grossberg.

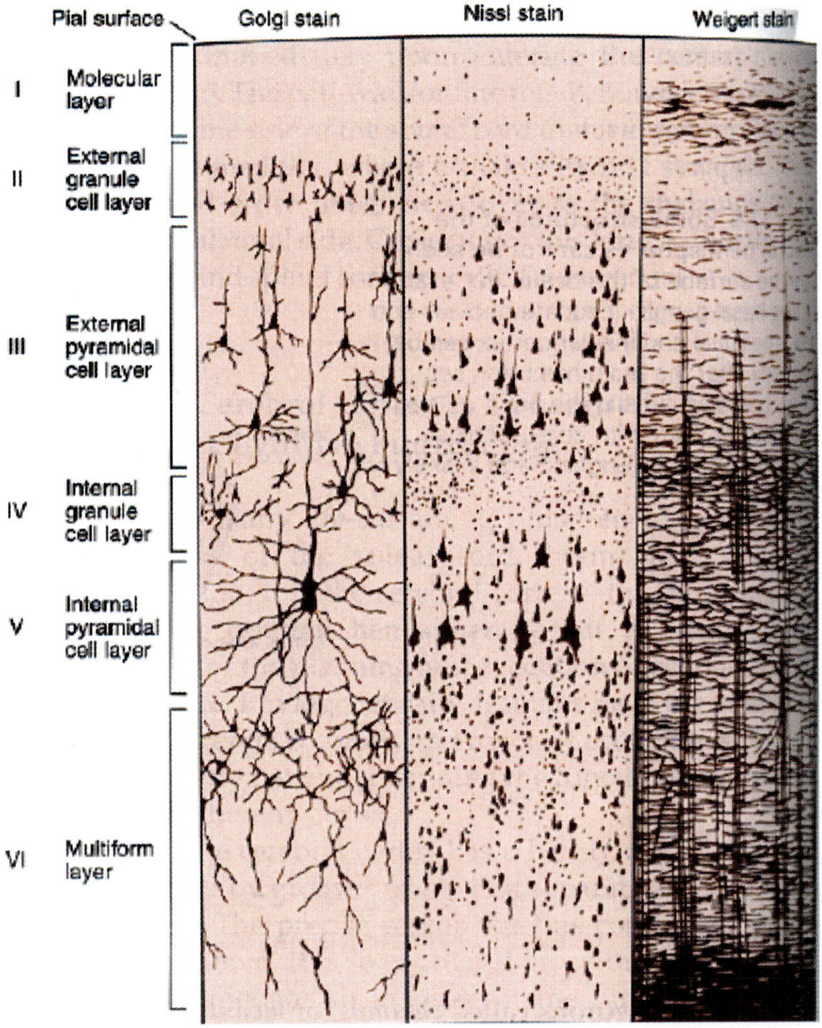

Fig. 7.5. Six distinct layers of the laminar architecture.

Top Down, Bottom-Up, and Horizontal Interactions

Evidence has surfaced in both anatomical and neurophysiological studies that the cells in the laminar layers have top-down (feedback), bottom-up (feedforward), and horizontal interactions. These have become one of the key design features in laminar computing. The fundamental question then is: "How are bottom-up, top-down, and horizontal interactions organized within cortical layers to generate adaptive behaviors?"

Brain signals related to the knowledge we have acquired about the world are called top-down. Signals related to incoming sensory information are called bottom-up. Grossberg and Carpenter have explained that bottom-up information by itself can activate the target nodes. Target nodes, in the biological sense, refer to the cells that represent the target. The top-down expectation selects consistent bottom-up signals, suppresses inconsistent bottom-up signals, and cannot by itself activate the target node. The role of top-down processing enables us to pay attention to learn expectations about the world.

Top-down influence can occur during face-to-face conversation — visual feedback (e.g. head and eye gesture) to communicate relevant information and to synchronize rhythm between participants. People often rely more on visual feedback than on visual perception.

It is noted that the connections "up" and "down" within the thickness of the cortex are more obvious and dramatically denser than the "horizontal" connections. Examples of horizontal neurons connection are shown below.

Figure 7.6 shows the horizontal neurons (A), horizontal pyramidal cells (B), and fan-shaped neurons (C) in layer VI (layer 6) of a 21-day old rat (Ferrer *et al.*, 1986). More information on horizontal neuron interactions can be found in the paper by (Novojilova and Babmindra, 2004).

Attention. Attention is a behavioral concept, but one whose properties arise from brain mechanisms. Attention can prime an observer to expect an object to occur at a given location or with particular stimulus properties (Duncan, 1984). According to Grossberg (Grossberg, 2005), attention is part of a unified design of top-down, bottom-up, and horizontal interactions among identified cells in laminar cortical circuits.

Attentive information processing is a fast processing loop, while learning and memory is a slower processing loop. Grossberg states

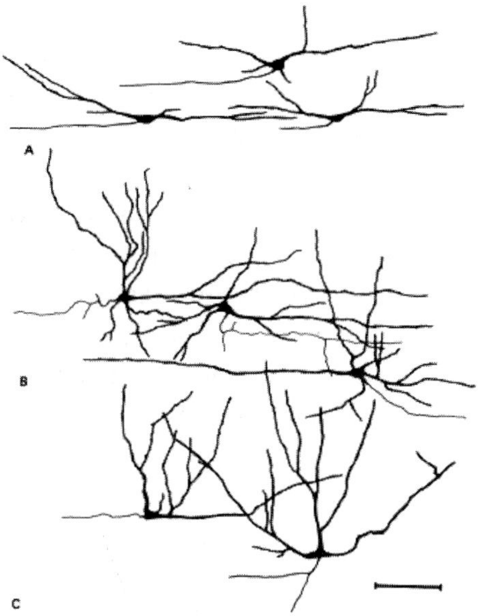

Fig. 7.6. Horizontal connecting neurons.

that top-down attentive feedback encodes learned expectations that self-stabilize learning in response to arbitrary temporal sequences of input spatial patterns in real time.

Attention is linked to learning. The role of attention in controlling adult plasticity and perceptual learning was demonstrated by (Ahissar and Hochstein, 1993) and (Gao and Suga, 1998). Gao and Suga reported physiological evidence that acoustic stimuli caused plastic changes in the inferior colliculus (IC) of bats only when the IC received top-down feedback from auditory cortex. These authors also reported that plasticity is enhanced when the auditory stimuli were made behaviorally relevant. Grossberg indicated that this is consistent with the ART[a] (Adaptive Resonance

[a] ART was introduced as a theory of human cognitive information processing (Grossberg, 1976), and later, the theory has evolved as a series of real-time neural network models that perform unsupervised and supervised learning, pattern recognition and prediction (Carpenter and Grossberg, 1987).

Theory) proposal that top-down feedback allows attended, and thus, relevant, stimuli to be learned while suppressing unattended irrelevant ones. Note: the evidence that cortical feedback also controls thalamic plasticity in the somatosensory system has been reported by (Krupa *et al.*, 1999).

Feedforward and feedback. There is rapid feedforward process when the bottom-up data is unambiguous. Feedback is automatically engaged, using top-down data, to choose between ambiguous alternatives.

When there is a good enough match between bottom-up and top-down signal patterns between two or more levels of processing, their positive feedback signals amplify and prolong their mutual activation, leading to a resonant state (Grossberg, 2003). The amplification and prolongation of the system's fast activations is sufficient to trigger learning in the more slowly varying adaptive weights that control the signal flow along pathways from cell to cell.

The neuron fires rapidly when stimuli is center focused with on-center (self-excitatory) and off-surround (competitive). If the stimuli is surround focused, the off-center and on-surround will cause rapid firing.

Figure 7.7 shows the recurrent (feedback) on-center off-surround network. This network is capable of contrast-enhancing its input pattern, normalizing its total activity, and storing the contrast-enhanced pattern in short-term memory (STM) without saturation (Grossberg, 1982).

Shunting Networks

The core mathematical framework for biophysically based neural modeling was developed by Sir Alan Hodgkin and Sir Andrew Huxley in the early 1950s. Their work won them a Nobel Prize. Hodgkin and Huxley used the giant squid's axon to carry out their experiment. They recorded the membrane potential dynamics of a single neuron and modeled them using a set of nonlinear

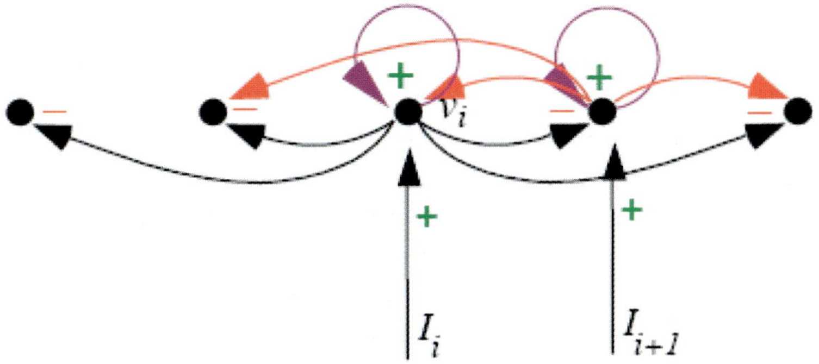

Fig 7.7. On-center off-surround network.

differential equations (Hodgkin and Huxley, 1952). See also (Nelson, 2004), which nicely summarized their work. Hodgkin and Huxley's cell membrane equation forms the basis of many subsequent studies on modeling the neurons.

Grossberg's shunting equation is a network variant of the original single cell Hodgkin and Huxley's cell membrane equation (see annex for the shunting equation and its correspondence to Hodgkin and Huxley's cell membrane equation).

However, some critics, such as (Douglas *et al.*, 1988) and (Ferster and Jagadeesh, 1992), question the "biological plausibility" of shunting in the cortex. These critics debated that shunting inhibition does not appear to be performed by cortical neurons.

However, to a physiologist, "shunting inhibition" is purely silent, meaning that if you stimulate an inhibitory fiber alone, no hyperpolarization of post-synaptic cells will occur. Biophysically, this means that the reversal potential of inhibition is at the resting potential, and more importantly, that large conductance changes occur.[b] The large conductance changes have a divisive or normalizing effect on

[b] Note that silent inhibition can also occur by just choosing the inhibitory saturation point to equal the resting potential. Thanks to the comment by Professor Stephen Grossberg.

excitatory current. Conductance change is equivalent to the change in the size of the denominator of the equilibrium solution of the shunting equation. So, the experimental data conducted by some scientists may have appeared to rule out significant inhibitory normalization via the denominator.

Experiments, such as (Heeger, 1992) and (Borg-Graham *et al.*, 1998), support the hypothesis that shunting inhibition does take place in cortical neurons. Hence, it is argued that though normalization clearly occurs in the cortex, a shunting network may not be the actual means, but at best is a simple and intuitive way of achieving the same effect.

Shunting network models are used by Grossberg and his colleagues in BU/CNS for modeling the neurons and their mutual interaction in the brain.

The shunting dynamics indicate that unexcited sites are "switched ON" by mass action from "their" (excitatory) inputs, and excited sites are "switched OFF" by mass action from "other" (inhibitory) inputs.

Folded Feedback

Another feature of the laminar computing is the folded feedback (another term coined by Grossberg). What is folded feedback? All cells in layer 2/3 send excitatory feedback signals to layer 6 via layer 5 (Gilbert and Wiesel, 1979; Ferster and Lindstrom, 1985). Layer 6, in turn, once again activates the on-center off-surround network from layer 6 to layer 4, which feeds into layer 2/3. Grossberg calls this process folded feedback (Grossberg, 1999). Note that cells in layer 2/3 are complex cells, and cells at layers 4 and 6 are simple cells.

To reiterate the idea of folded feedback, consider the attentional signals from the higher cortex being fed back to layer 4 of V1, via the modulatory 6 to 4 path. Corticocortical feedback axons tend preferentially to originate in layer 6 of the higher area and to terminate in layer 1 of the lower cortex, where they can excite the apical dendrites of layer 5 pyramidal cells whose axons send collaterals into layer 6. Having arrived in layer 6, the feedback is then "folded" back up into the feedforward stream by passing through layer 6 to 4.

Grossberg explained that when ambiguous or complex scenes are being processed, intracortical folded feedback enables stronger groupings that started to form in layer 2/3 to inhibit weaker groupings, whereas intercortical folded feedback enables higher-order processing constraints to bias which groupings will be selected. Grossberg further explained that both infants and adults can focus their attention selectivity upon whole objects because of attention causing an excitatory modulatory bias at some cells in layer 4, and groupings that form in layer 2/3 can be enhanced by this modulation via their positive feedback loops from 2/3 to 6 to 4 to 2/3. Grossberg's employs this folded feedback to explain how visual attention can flow along an object boundary and also an illusory contour (Roelfsema *et al.*, 1998; Raizada and Grossberg, 2001). See also the LAMINART model at later sections.

Figure 7.8 shows the intracortical feedback from layer 2/3 to layer 6 and intercortical feedback from the higher regions to layer 6 of the lower regions. This intracortical feedback inhibits weaker groupings and enables stronger groupings to form. The intercortical feedback enables higher order processing, such as attention to bias which groupings will be selected. The intercortical attention acts via a top-down modulatory on-center off-surround network, and the intracortical loop acts to stabilize learning.

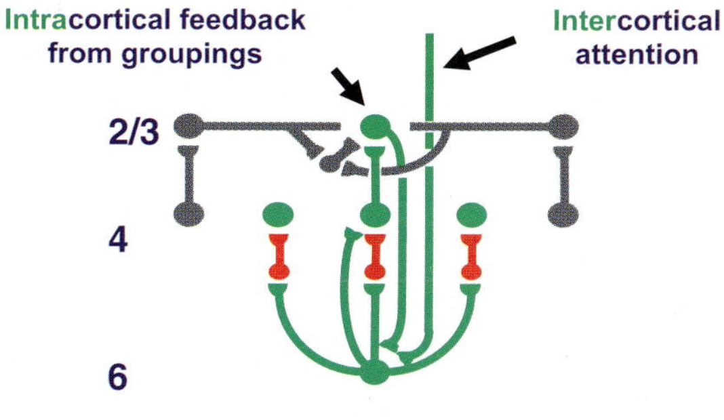

Fig. 7.8. Intracortical and intercortical feedback connection in the laminar layers.

The intracortical and intercortical feedback circuits that control the property of attentional selection have been shown in modeling studies to play a key role in stabilizing infant development and adult perceptual learning within multiple cortical areas, including cortical areas in V1 and V2.

Complementary Properties

In (Grossberg, 2000) paper, Grossberg presented evidence that the brain's processing streams compute complementary properties. Each stream's properties are related to those of a complementary stream much as a lock fits its key, or two pieces of a puzzle fit together. Pairs of parallel cortical processing streams compute complementary properties in the brain. Grossberg illustrated the complementary properties using the ventral ("what") and dorsal ("where") streams of the visual pathway. See Fig. 7.9 below for the two streams in visual pathway.

What and where stream. Fig. 7.9 shows some visual processes and their anatomical substrates that are being modeled as part of a unified vision system. LGN = Lateral Geniculate Nucleus; V1 = striate visual cortex; V2, V4, MT, MST = prestriate visual cortex; IT = Inferotemporal cortex; PPC = posterior parietal cortex; PFC = prefrontal cortex. Understanding how the brain sees is one of the areas where experimental and modeling work have advanced the furthest, and this progress illustrates several different types of complementary interactions. The figure provides a schematic macrocircuit of the types of processes that are assembled into a unified theory of how the brain sees, including processes of vision, recognition, navigation, and cognition. In particular, key matching and learning processes within the "what" and "where" cortical streams have been proposed to be complementary: The "what" stream, through cortical areas V1-V2-V4-IT-PFC, learns to recognize what objects and events occur. The "where" stream, through cortical areas V1-MT-MST-PPC-PFC, spatially localizes where they are, and acts upon them. Complementary processes also occur with each

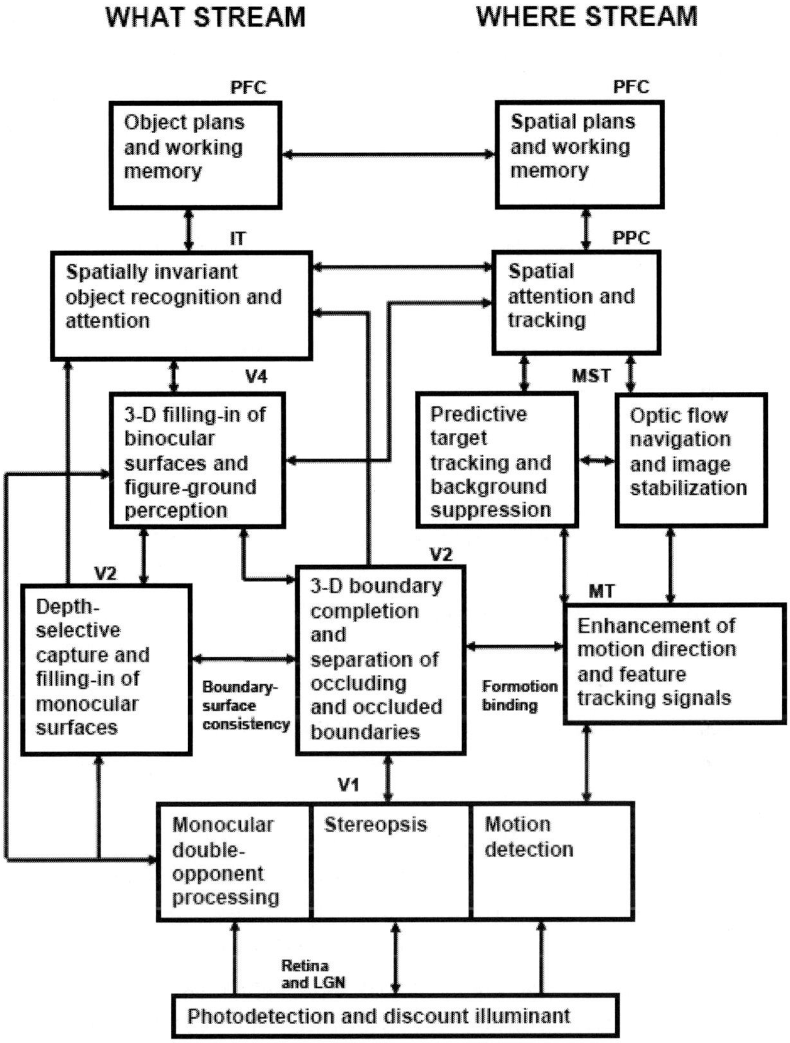

Fig. 7.9. A proposed unified vision system.

stream: the "what" stream boundary grouping via the (V1 interblob)-(V2 pale stripe)-V4 stages, and surface formation via the (V1 blob)-(V2 thin stripe)-V4 stages, have complementary properties. The "where" stream target tracking via MT-(MST ventral) and navigation via MT-(MST dorsal) have complementary properties. Such complementary processes are predicted to arise from symmetry-breaking operations during cortical development (Grossberg, 2003).

The "Drum-up" to Laminar Computing

From 1958, Stephen Grossberg, as an undergraduate student, started to derive neural network learning equations. Later, as a PhD student in Rockefeller University, he conducted research on nonlinear dynamics learning equations. During the late 1960s and early 1970s, he continued to study the nonlinear dynamics in the context of how network of cells work in a recurrent on-center off-surround anatomy.

From recurrent on-center off-surround network to shunting network. In the late 1960s, experimental studies suggested that cells are arranged in a recurrent on-center off-surround anatomy (Anderson *et al.*, 1969; Stefanis, 1969) (Eccles *et al.*, 1967). Based on this fundamental principle, Grossberg built his system equation called the recurrent on-center off-surround networks (Grossberg, 1973). (Details of the equation are given in the notes.) The recurrent on-center off-surround network equation is shown to resolve the noise-saturation dilemma in neural networks. Grossberg indicated that the noise-saturation dilemma confronts all noisy cellular systems that process input patterns. If the inputs are too small, they can get lost in noise. If the inputs are too large, they can turn on all excitable sites, thereby saturating the system and rendering it insensitive to input differences across the cells. Grossberg explained that the competitive interactions among the cells from the recurrent on-center off-surround networks automatically retune their sensitivity to overcome the saturation problem (Grossberg, 1980). One of the key properties of the recurrent on-center off-surround networks is the normalization property or shunting effect. Grossberg indicated that in the neural context, the competitive interactions are shunting interactions, and they are carried out within on-center off-surround anatomy. The off-surround signals automatically inhibit or shunt the networks, i.e. they self-normalize the networks total activities.

Hence, the recurrent on-center off-surround network was later also known as the shunting on-center off-surround network (Ellias and Grossberg, 1975) or shunting cooperative-competitive feedback networks (Grossberg, 1982). It was then reformulated as the current shunting networks and used in the short-term memory (STM) model and many other experiments related to vision process modeling (Grossberg, 1988).

From shunting network to solving the stability-plasticity dilemma. In the 1980s and 1990s, Grossberg also discussed the issue of stability and plasticity dilemma in the brain. Grossberg called the problem — whereby the brain learns quickly and stabily without catastrophically forgetting its past knowledge — the stability-plasticity dilemma. Grossberg indicated that the stability-plasticity dilemma must be solved by every brain system that needs to rapidly and adaptively respond to the flood of signals that subserves even the most ordinary experiences. If the brain's design is parsimonious, then similar design principles operating in all the brain systems that can stabily learn an accumulating knowledge base in response to changing conditions throughout life (Grossberg, 1999, 2003).

Grossberg suggested that top-down attention is a key mechanism whereby the brain solves the stability-plasticity dilemma. Grossberg explained that, without an appropriate type of nonlinear top-down learned feedback, one cannot escape this instability in response to a non-stationary input environment. This suggests a simple type of top-down on-center off-surround feedback circuit (i.e. shunting networks) would provide additional computational power to solve the problem. As discussed earlier, the shunting network has inherited a nonlinear feedback process and can generate basic properties like noise suppression, automatic gain-control normalization, and stable learning.

The link-up with ART. In early 1970s, Grossberg explained that the functional unit of perception and cognition is a state of resonant activity within the whole system. And the resonant state can drive adaptive or learned changes in the system structure. The resonant state is therefore called an adaptive resonance. Adaptive resonance arises when feedforward (bottom-up) and feedback (top-down) signals within the systems are consonant. The feedback computation corresponds to our intuitive notion of expectancies. Feedback expectancies help to stabilize the code against erosive effects of irrelevant environmental fluctuations.

Adaptive resonance theory (ART) was introduced in 1976. Grossberg explained that ART helps to self-stabilize learning in a real-time environment where learning needs to remain stable through time, free from catastrophic forgetting, particularly when the amount of data becomes large and can change its statistical properties through time (Grossberg, 1976).

ART was introduced as a theory of human cognitive information processing (Grossberg, 1976), and later, the theory evolved into a series of real-time neural network models that perform unsupervised and supervised learning, pattern recognition and prediction (Carpenter and Grossberg, 1987) (Carpenter and Grossberg, 2003). A series of ART-based models was proposed in the last 20 years of research in BU/CNS. These include CogEM, START, iSTART, etc. (See notes for summaries of the various ART-based models.)

From ART to LAMINART in vision. Why the LAMINART model? Grossberg explained that laminar layers follow the adaptive resonance theory (ART), or ART could be realized within identified laminar cortical circuits. ART is a perceptual and cognitive theory that proposes how stable development and learning can occur throughout life using top-down attention. ART predicted in the 1970s that attention can be thought of as a top-down modulatory on-center off-surround network.

The LAMINART model proposes how bottom-up, top-down, and horizontal cortical circuits work together in laminar circuits and how they realize processes of development, learning, grouping, and attention.

The LAMINART model also explained how the visual cortex organizes for processing boundary and surface information, how preattentive and attentive processing are intimately linked within the cortical layers, and how laminar computing enables the visual cortex to realize: (1) the binding process whereby cortex groups distribute data into coherent object representations; (2) the attentional process whereby the cortex selectively processes important events; and (3) the developmental and learning processes whereby the cortex shapes its circuits to match environmental constraints. It is proposed that the mechanisms governing (3) in the infant lead to properties (1) and (2) in adults.

Figure 7.10 shows a LAMINART model. The model is a synthesis of feedforward (bottom-up), feedback (top-down), and horizontal interactions within and between the lateral geniculate nucleus (LGN) and visual cortical areas V1 and V2. Cells and connections with open symbols indicate excitatory interactions, and closed symbols indicate inhibitory interactions. The triple top-down connections indicate attention feedback (Grossberg, 2000b; Grossberg, 1999).

The LAMINART model is used to simulate neurophysiological, anatomical, and psychophysical data about the visual cortex. Variants of the model are finding their way into new algorithms for processing complex and noisy imagery with a large dynamical range such as SAR imagery.

The LAMINART model has the property of hybrid feedforward and feedback computing. When an unambiguous scene is processed, the model can quickly group the scene in a fast feedforward sweep of activation that passes directly through layer 4 to 2/3, and then on to layers 4 to 2/3 in subsequent cortical areas. This property clarifies how recognition can be fast in response to unambiguous scenes. If, however, there are multiple possible groupings; say, in response to a complex textured scene, then competition among these possibilities due to inhibitory interactions in layers 4 and 2/3 can cause all cell activities to become smaller. This happens because the competitive circuits in the model are self-normalizing; that is,

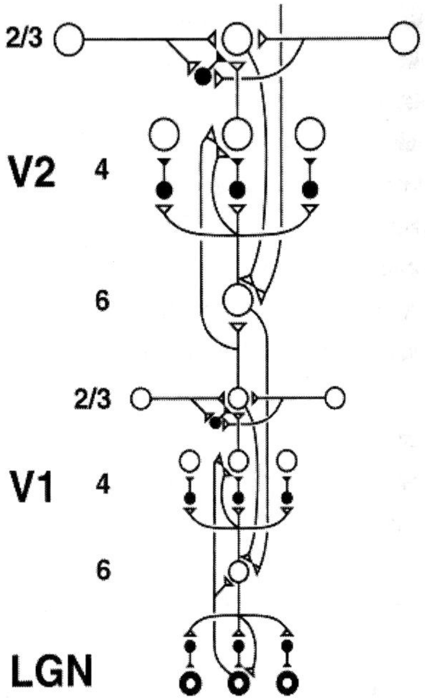

Fig. 7.10. A LAMINART model.

they tend to conserve the total activity of the circuit. This self-normalizing property emerges from shunting on-center off-surround networks that process input contrast over a large dynamic range without saturation (Grossberg, 1973; Douglas *et al.*, 1995).

Grossberg indicated that these self-normalizing circuits carry out a type of real-time probability theory in which the amplitude of cell activity covaries with the certainty of the network's selection, or decision, about a grouping. Amplitude also covaries with processing speed. Low activation greatly slows down the feedforward processing in the circuit because it takes longer for cell activities to exceed output thresholds and to activate subsequent cells above the threshold.

In the LAMINART model, network uncertainty is resolved through feedback, i.e. weakly active layer 2/3 grouping cells feed back signals to layers 6, then 4, then 2/3, to close a cortical feedback loop that rapidly contrast enhances and amplifies a winning grouping. As the winner is selected and weaker groupings are suppressed, its cells become more active, and hence, can again more rapidly exceed output thresholds and send the cortical decision to subsequent processing stages. In summary, the LAMINART circuit behaves like a real-time probabilistic decision circuit that operates in a fast feedforward mode when there is little uncertainty, and automatically switches to a slower feedback mode when there is significant uncertainty. Feedback selects a winning decision that enables the circuit to speed up again. Activation amplitude and processing speed both increase with certainty. The large activation amplitude of a winning grouping is facilitated by the synchronization that occurs as the winning grouping is selected.

The LAMINART model clarifies how excitatory and inhibitory connections in the cortex can develop in a stable way by achieving and maintaining a balance between excitation and inhibition. Long-range excitatory horizontal connections between pyramidal cells in layer 2/3 of the visual cortical areas play an important role in perceptual grouping (Hirsch and Gilbert, 1991; McGuire *et al.*, 1991). The LAMINART model proposes how development enables the strength of long-range excitatory horizontal signals to become balanced against that of inhibitory signals, which are mediated by short-range di-synaptic inhibitory interneurons that target the same target pyramidal cells.

Variants of LAMINART. A 3-D LAMINART model was proposed in 2003.

Figure 7.11 shows a 3-D LAMINART model (Grossberg and Howe, 2003), including 3-D boundary completion and attention, as well as the binocular and monocular interactions. Due to the binocular fusion that occurs in layer 3B, the binocular boundaries that are formed in layer 3B and 2/3A can be positionally displaced or shifted, relative to their monocular input signals, to layers 6 and 4. The model proposed that horizontal connections occur in layer 5. Feedback signals from layer 2/3A propagate vertically to layer 5, whose cells activate horizontal axons in layer 5. In 2007, the dARTEX neural model was proposed, illustrating how laminar interactions in the visual cortex may learn and recognize object

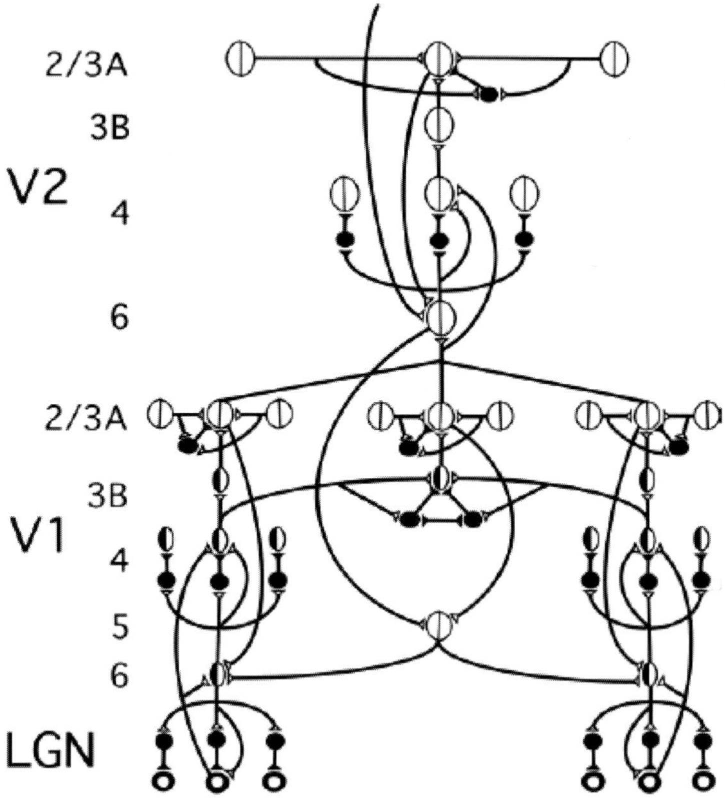

Fig. 7.11. A 3-D LAMINART model.

texture and form boundaries. The model combines multiple-scale bottom-up filtering, horizontal grouping, top-down spatial and object attention, and a distributed ART (dART) classifier, in a laminar cortical circuit model. dARTEX is an extension of ARTEX. The model unifies five interacting processes: region-based texture classification, contour-based boundary grouping, surface filling-in, spatial attention, and object attention. dARTEX shows how form boundaries can determine regions in which surface filling-in occurs; how surface filling-in interacts with spatial attention to generate a form-fitting distribution of spatial attention or attentional shroud;

how the strongest shroud can inhibit weaker shrouds, and how the winning shroud regulates learning of texture categories, and thus the allocation of object attention (Bhatt *et al.*, 2007).

From Vision to Cognition

The vision process activates about 50% of the brain cerebral cortex, and hence, most work has been devoted to vision research. However, increasingly the focus has been moving toward cognition. Can Grossberg's LAMINART principles be used to explain data on the temporal dynamics of cognitive information processing? In particular, how do the layered circuits of the prefrontal and motor cortex carry out working memory storage, sequence learning, and voluntary sequential performance? A neural model called LIST PARSE (Grossberg and Pearson, 2006) has begun to explain and quantitatively simulate cognitive data about immediate serial recall and free recall, including bowing of the serial position performance curves, error-type distributions, temporal limitations upon recall accuracy, and list length effects. (Note: bowing means that the system is not able to reproduce the correct order of items from working memory if the list of stored items becomes too long. Thus the inability to read-out correct order information of long lists from working memory can be traced to the need for stable learning.) The model also qualitatively explains cognitive effects related to attention, temporal grouping, variable presentation rates, phonemic similarity, presentation of non-words, word frequency/item familiarity and list strength, distracters and modality effects. In addition, the model quantitatively simulates neurophysiological data from the macaque prefrontal cortex obtained during sequential sensory-motor imitation and planned performance demonstrating parallel coding of movements in a sequence with relative activation predictive of performance order (Grossberg, 2006).

Grossberg indicated that the LIST PARSE model proposes functional roles for the different layers of granular lateral prefrontal cortex, notably, the ventrolateral prefrontal cortex, for storage of temporal lists of events in working memory and learning of list categories. LIST PARSE illustrates how variations on granular laminar cortical circuits can quantitatively simulate data about spatio-temporal processes in vision as well as spatio-temporal data about cognition.

Notes: This provides a quick glance through Professor Stephen Grossberg's work. All the materials in this chapter are largely summaries and extracts from the published work of Grossberg, Carpenter, and their colleagues.

Chapter 8

PROBABILISTIC COMPUTING: THE BAYESIAN MIND

Within the neural modeling community, one choice of mathematical tools uses Bayesian principles and graphical models. This mathematical approach is chosen due to its claim of clear semantics and an expressive medium for capturing neural function at multiple levels of details. For example, in modeling the visual hierarchical cortex architecture, a number of scientists proposed a Bayesian network approach to model this hierarchical architecture.

As early as 1991, Mumford had started to model the brain's cognitive process using the Bayesian method (Mumford, 1991; Mumford, 1992). In (Lee and Mumford, 2003), they extended their hierarchical Bayesian model to account for processing at multiple regions in the temporal cortex including V1, V2, V4 and the inferotemporal cortex (IT). They assumed that each of these regions is responsible for computing features at different levels of abstraction. And (Dean, 2005) further generalized their model to include an account on the hierarchical layer of the cortex.

Subsequently, the hierarchical Bayesian model was extended to include belief propagation (Isard, 2003; Sudderth *et al.*, 2003; Rao and Ballard, 1996; Zernel, 2000; George and Hawkin, 2005).

Is the brain operating on Bayes rules? Some scientists have a strong view that it is not, and others think that Bayes rules and Bayesian inference have a role in the brain. From the behavioral point of view there are claims that the brain exhibits some probabilistic "likeness".

Furthermore, some believe that behavior drives the brain rather than the brain drives behavior.

I believe the Bayesian approach is useful in the sense that it enables us to explicitly explore the role of prior knowledge and combine evidence of the likelihood of events. If we take the notion that the brain is an information processing machine, then information processing will typically involve inferring new information from information that has been derived from the senses or from the memory. The process of inferring typically needs to take into account uncertainty. Probabilistic methods such as Bayesian techniques can explicitly take into account uncertainty. Besides, Bayesian modeling can serve as a good approximation to address high-level cognition and new data can be used to update existing estimates of the most likely model of the world. The use of Bayesian methods for inference in cognition have been used by a number of cognitive scientists (Chater *et al.*, 2006a,b; Gopnik *et al.*, 2004; Gopnik and Tenenbaum, 2007). In cognitive psychology, intelligent actions are thought to come about due to causal processes and mechanisms. And causal Bayesian networks are one of the graphical methods for representing the causal construction (Glymour, 2001).

> "*Connectionism will be affected by the increasing appeal to Bayesian probability theory in human reasoning.*" — (Thomas and McClelland, 2008).

(McClelland and Thompson, 2007) reported that work has already begun to relate connectionist and Bayesian accounts and is used in the domain of causal reasoning in children. In some cases, connectionism may offer alternative explanations of the same behavior; in others it may be viewed as an implementation of a Bayesian account.

However, the Bayesian method may be difficult to scale up, i.e. for a large scale information processing problem, the technique may have limitations unless a good approach can be used to overcome the limitations in the future.

In the following, we present one approximate approach of modeling the human cognitive process using the Bayesian method.

Probabilistic Model of Cognitive Process

In (Ng *et al.*, 2006, 2007), a cognition-based dynamic reasoning machine called D'Brain was developed. D'Brain stands for **D**ynamic **B**ayesian **R**easoning and **A**dvanced **I**ntelligent **N**etwork. D'Brain is a high level cognitive system designed to provide decision support to a human decision-maker. The design of D'Brain's architecture incorporated a cognitive perspective, so it has some level of resemblance to the human cognitive process, which is a key element in building the next generation of truly intelligent cognitive aids for humans (Ng, 2003).

In designing D'Brain's cognitive framework, analogies were drawn between the framework components and the human brain components. The brain regions important for cognitive functions include the neocortex (the seat of higher intelligence), the limbic system, such as the hippocampus, which is important for memory formation, and the thalamus that plays a critical role for relaying information from the sensory to diverse brain regions. Note that in this work, we loosely equate the thalamus to a resource manager.

Memory is an important component in the study of cognitive intelligence (Ng, 2003). The hippocampus has an important role in the formation of new memories about experienced events (episodic memory) and is part of a larger limbic system involved in learning and the consolidation of long-term memories. There are also the sensory memories and working memories in the brain.

Figure 8.1 shows a very simplified overview of the brain.[a]

In designing the D'Brain cognitive framework, various cognitive mechanisms proposed by (Endsley, 2000), which are important for the development of situation awareness, have been considered. D'Brain adopted Endsley's concept of instantiating

[a] Figure 8.1 is by no means completed nor does it provide sufficient details for the complete implementation of the entire brain.

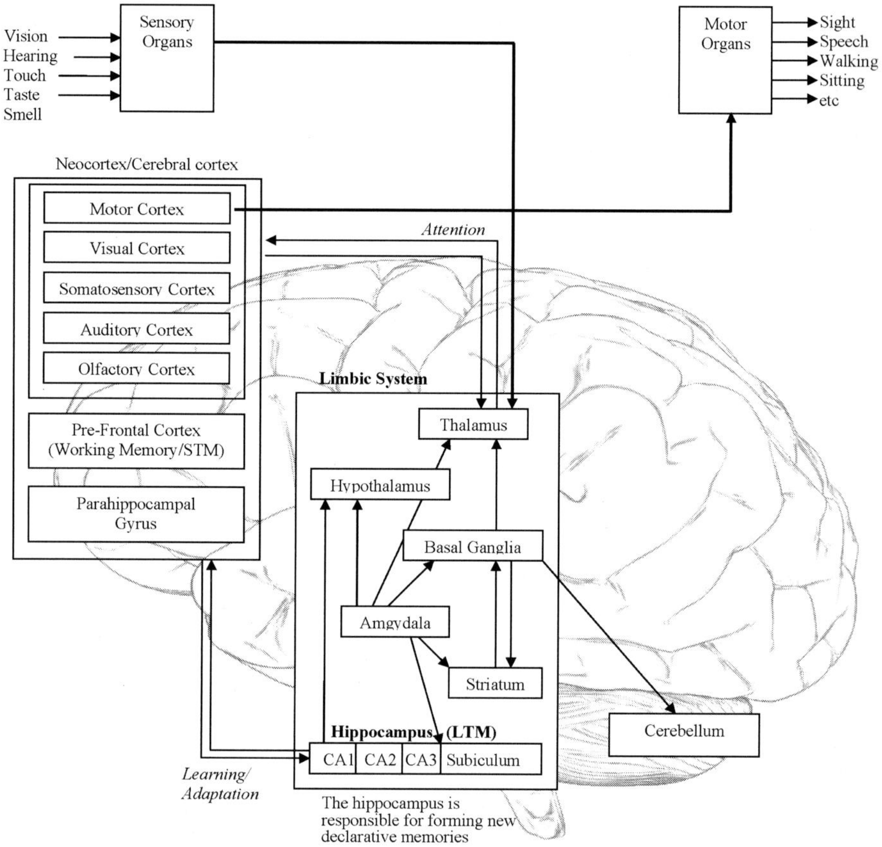

Fig. 8.1. A simplified architecture of the human brain system.

situation models from mental models for reasoning. However, adopting from the human brain model, the D'Brain design improved on Endsley's model by introducing the learning/adaptation module.

Mental models are mechanisms whereby humans are able to generate descriptions of system purpose and form, explanations of system functioning and observed system states, and predictions of future states. Mental models are used to describe a person's representation of some systems of interest. Mental models embody stored long-term knowledge about these systems that can be called upon to direct problem solving and interaction with other relevant systems when needed. In a way, we can think of mental models as stores of long-term memory. On the other hand, a situation model is a schema depicting the current state of the mental model of the system, and hence, is stored in the short-term memory. As new observations of the changing environment are made, the situation model is updated. Based on information of the current environment, the situation model draws knowledge from relevant mental models in order to explain the current situation. In this sense, the situation model is the combination of instantiations of various mental models. In summary, mental models represent general knowledge and are static and more generic, whereas a situation model is specific to the current situation, and hence, more dynamic. In particular, a situation model not only captures the human's representations of the various parameters of the system, but also includes an understanding of the dynamics of the system developed from changes in the situation model over time. While a situation model is developed from observing the current world, it is also largely influenced by the underlying mental models that the person has accumulated through learning from past experiences in other worlds. By reasoning using situation and mental models, we can determine what information should be attended to, how that information should be interpreted and integrated, and what projections can be made about what will happen to the system in the near future (perception, comprehension, and projection). Figure 8.2 shows the design of the cognitive framework for an application on activity and intent inference.

The cognitive framework for activity and intent inference draws some analogy with the biological components of our brain. For example, the Intent Inference module and Model Matching model of the framework are equated to components of the neocortex. The Attention module and Context/Environment Information Management module are equated loosely to components of the thalamus. Mental models are like factual knowledge stored in our long-term declarative semantic memory. Situation-specific intent models are like events stored in our long-term declarative episodic memory. In our cognitive

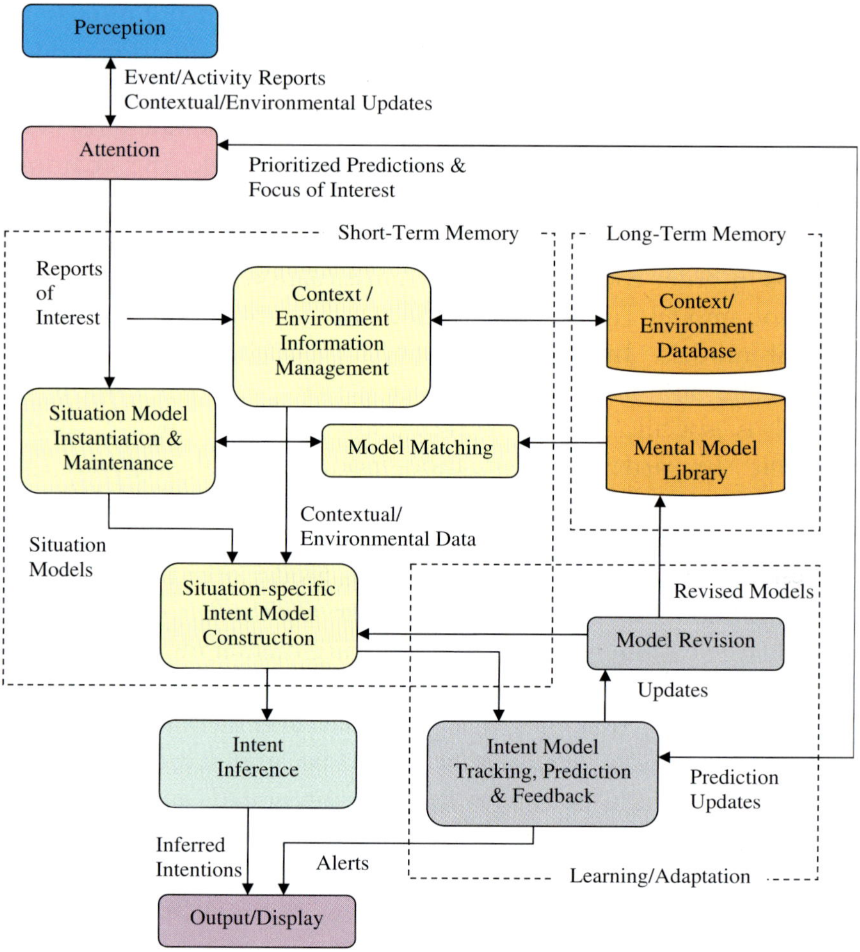

Fig. 8.2. Cognitive framework for activity and intent inference.

framework, mental models are dynamically pieced together in real-time as the external environment changes, and hence, our interpretation of the situation is updated. This is similar to how our semantic memory of our brain, which stores bits of factual information, are pieced together to interpret a certain event/episode. In our Intent Model Tracking, Prediction and Feedback Module of our framework, situation-specific intent models are stored and tracked. This memory used for storing situation-specific intent models is similar to our declarative episodic memory formed by our hippocampus. The Model Revision module is then similar to how we generalize several monitored episodic memory to obtain new factual knowledge, which is then incorporated into semantic memory. The ability to store and track episodes (the function of the hippocampus) is important for learning new knowledge. The algorithms used to select which models to combine and how to combine them can be said to be akin to cognitive skills. The memory that stores these algorithms is an analogy to our brain's long-term procedural memory, which also stores skills.

An advantage of our framework design is that the implementation of mental models, and hence, situation and intent model, is not restricted to a particular knowledge representation. We can use Bayesian networks, neural networks (Haykins, 1999), fuzzy rules (Zadeh, 1974), templates, etc. Furthermore, we can have a mixture of such mental model representations and fuse them during the intent model construction process by, say, the Dempster–Shafer algorithm. However, the choice of the mental model representation has a major impact on the implementation of the inference system. It affects the way in which the intent model can be constructed through combination of the situation models. Nonetheless, the problem of using a new knowledge representation might be tackled by adding new algorithms to our declarative memory. The new algorithms can be added manually, or ultimately, the system might be able to learn new algorithms autonomously. We acknowledge that combining various mental model representations, such as Bayesian networks and neural networks, is not a trivial task. However, we point out that our architecture does not limit us to a single mental model representation such as Bayesian networks. The advantage our architecture has is that it has the flexibility to combine various mental models that are constructed in different representations to form a situation model. For example, a knowledge engineer well-versed in rule-based systems can construct a set of knowledge fragments in rules, while another knowledge engineer well-versed in probabilistic models can construct another set of knowledge fragments in Bayesian networks and the architecture allows the fusion of the knowledge fragments from these two experts.

Bayesian Networks

For our first implemented system, we chose to use Bayesian networks because a strong analogy exists between our cognitive framework and the knowledge-based approach using Bayesian networks. Furthermore, we take the view of the brain being as an information processor, and information processing typically involves inferring new information from information that has been derived from the senses or memory. The process of inferring typically needs to take into account uncertainty, and hence, the field of probabilistic methods is considered.

The concept of building situation-specific Bayesian networks (SSBN) (Wright *et al.*, 2002; Laskey *et al.*, 2001) from a library of network fragments coincides with the idea of instantiating situation models from mental models (network fragments are the implementation of long-term memory (LTM). This includes "context/environment database" and "mental model library" modules in our cognitive architecture). Building SSBN corresponds to "model matching", "context/environment information management", "situation model instantiation and maintenance", and "situation-specific intent model construction" modules in our architecture. The fragment library is viewed as long-term memory that embodies static knowledge of different domains and the situation specific network as the current working memory or short-term memory (STM).

Most of the solutions reported in publications on using Bayesian networks for inference are based on the concept of building a complete and static Bayesian network for carrying out the inference process. The disadvantage of such static modeling is that the events and activities offer changes from what have been expected. Hence, the static network will not be able to fully represent the new situation or adapt sufficiently well to it. To adapt to dynamic situations, instead of using a static Bayesian network to reason about the situation, we use an evolving situation-specific Bayesian network (SSBN) that is formed by fusing Bayesian network knowledge fragments together according to the sensor inputs received. We build our mental models of a certain domain by first constructing these Bayesian network knowledge fragments. These fragments form a knowledge base of the domain. Each fragment models a Bayesian network with a network structure that generally holds true, and each fragment behaves like a standard Bayesian network. As evidence is observed, knowledge (or network) fragments are fused together to form a SSBN (Fig. 8.3). In other words, we maintain a library of sub-networks fragments that encode a smaller set of relationships that generally hold true. These fragments are then fused at run-time based on the presence of evidence, and is constantly updated to reflect the current situation. When evidence streams in, for example, when node A is observed (Fig. 8.3), the system will search for fragments that contain node A and fuse them together with the previous SSBN to form an updated and

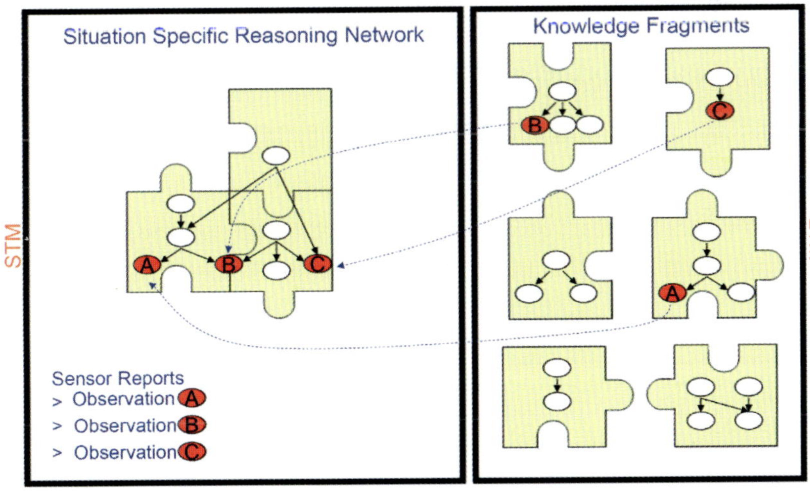

Fig. 8.3. Dynamic reasoning using knowledge fragments.

current SSBN. Likewise, when node B is observed, the system will fuse fragments that contain node B with the previous SSBN to form the new and current SSBN. We also acknowledge that the system does not need to fuse all fragments that contain the observed node. There needs to be some control in order to control which fragments will be fused into the SSBN. Also, fragments can be deleted when they are no longer relevant to the current situation.

Knowledge bases are akin to libraries and the network fragments belonging to a knowledge base are akin to the books that belong to a particular library. Note that each knowledge fragment is itself a standard Bayesian network (Pearl, 1988).

The equations for fusing Bayesian network knowledge fragments to create the SSBN in order to perform dynamic reasoning can be found in the Notes for this chapter.

Conclusions

This chapter presented the perspective of modeling the brain using the probabilistic approach. The advantages and disadvantages of using this probabilistic approach were discussed. An example of a probabilistic model for the cognitive process was presented. The probabilistic model using Bayesian networks and its modeling of the "long-term memory" and "short-term working memory" were discussed.

Chapter 9

THINKING MACHINE: HIGHER THEORIES OF BRAIN AND COMMONSENSE KNOWLEDGE GENERATION

"*You can't think about thinking, without thinking about thinking about something.*" — Seymour Papert (Papert, 2005).

One clear distinct feature humans have is the ability to think. Every moment of our lives, we are thinking. We are so involved in using our brain for thinking that sometimes we do not even notice it.

"*What makes us such excellent animals? We do not have the strength of an ox, an antelope's speed, or the grace of a cat. However, we humans are unsurpassed in our flair for developing new forms of art. We fabricate weapons, garments and dwellings; we are matchless at all sorts of social inventions; we make codes of behavior, with laws to enforce them — and then invent clever new ways to evade them. And surely our most outstanding trait is our knack for inventing new Ways to Think.*" — Marvin Minsky, The Emotion Machine.

"*A test of true thinking must involve emotion.*" — Rosalind Picard, Affective Computing.

Professor Howard Gardner, in his book, "Frames of Mind", indicated that human intelligence consists of multiple forms including social intelligence, which consists of interpersonal and intrapersonal skills. Peter Salovey and John Mayer identified these latter skills as emotional intelligence, which they defined as "*the ability to monitor one's own and others' feelings and emotions, to discriminate among them and to use this information to guide one's thinking and actions*" (Salvoey and Mayer, 1990).

What constitutes thinking? To "think", as defined by Webster dictionary is: "to have in mind", "to reflect", to ponder", "to call to mind"; "remember", "reason", "form a mental picture; imagine". It is also clear from normal human cognition that thinking and feelings are partners. If we wish to design a computer that "thinks" like a human brain, then it must also "feel". For example, to pass the Turing Test,[a] one may ask a question that involves a tragic accident. The details of emotion and cognition are discussed in Chapter 6.

Hence, to be able to think, one needs to have intelligence, emotion, and be in a conscious state. The conscious state here refers to the awake state where the brain is in a normal operating state or is functioning normally.

As "thinking" is a pretty high level concept, I will discuss some of the high level theories of the brain's machinery proposed by scientists. The high level theories may potentially explain how we think and how we can make a machine "think" like a human. These include:

- Multiple layers of machinery by Marvin Minsky
- Theory of Multiple Intelligences by Howard Gardner
- Triarchic theory of intelligence by Robert J. Sternberg
- Conceptual Blending Theory by Gilles Fauconnier and Mark Turner

[a] The Turing test was first proposed by Alan Turing in 1950s. It is a test to determine if the computer can demonstrate human-level intelligence and pass the test by convincing the human judges of their humanness.

The ability to think and behave intelligently requires both large quantities of common sense knowledge and the ability to learn from experience. The brain has a huge amount of internal knowledge stored, and more so commonsense knowledge. Psychologists studying young children's brains know that children as young as six years old already have a huge amount of commonsense knowledge. How could a machine be able to have such commonsense knowledge? Today, we are not able to model a biological way of representing commonsense knowledge. A number of current attempts use more conventional approaches. These include Cyc, WorldNet, and Open Mind (Open Mind Common Sense), and they will be discussed in this chapter.

Human thinking takes place in a cumulative process of growth and development and perhaps the ultimate process in thinking is to get wisdom. In the last section of this chapter, I include some discussion on what is wisdom.

We begin the chapter by discussing the point: Can a machine think and have a mind?

Can a Machine Think and Have a Mind?

> "*What if... we were magically shrunk and put into someone's brain while he was thinking. We would see all the pumps, pistons, gears, and levers working away, and we would be able to describe their working completely, in mechanical terms, thereby completely, describing the thought processes of the brain. But that description would nowhere contain any mention of thought! It would contain nothing but descriptions of pumps, pistons, levers!*" — Gottfried Leibniz 1690.

> "'*Could a machine think?' On the argument advanced here only a machine could think, and only very special kinds of machines, namely brains and machines with internal causal powers equivalent to those of brains. And that is why strong AI*[b] *has little to tell us about thinking,*

[b] Strong AI refers to artificial intelligence that matches or exceeds human intelligence. Strong AI is a term used by futurists, science fiction writers, and some researchers.

> *since it is not about machines but about programs, and no program by itself is sufficient for thinking.*" — John Searle, Minds, Brains, Program (Searle, 1980).

Computer systems that solely take instructions from software programs and execute rules as instructed are not thinking. In some sense, we are not programmed in this way and we have yet to discover the way the human brain produces thought. However, if one day we could design a "program", so to speak, which is dynamic and self-evolving, and give it the same plasticity as brain neurons and synapses, then maybe we would able to design a computer that can think on its own.

Currently, some programs are demonstrated to have simple adaptations, such as the change of parameter values due to specific input. These include programs such as artificial neural networks, genetic algorithms, recursive least square types of algorithms, and many others. However, we are far from achieving highly "plastic" software programs or equivalent forms in hardware that match the plasticity and stability of our brain's neurons and synapses.

A number of AI and cognitive psychologists have proposed a number of high level theories to explain the way human brain thinks and how to emulate it. Some of these schemes will be discussed in the next sections.

Multiple Layers of Machinery

What do we mean when we say "I am conscious and thinking?" Are there a number of smaller machineries in our brain that represent this? Professor Marvin Minsky, MIT Media Laboratory, thinks that the human brain has many different types of machinery or agents working together. Minsky thinks that no single theory can explain how the brain works. The human brain probably has many axioms and many small facts stored away as we learn. Minsky described a few high level theories that discussed how the brain works and how it can lead to a thinking machine (Minsky, 2006).

Self-reflective system. Minsky and Singh discussed designing a self-reflective system (Minsky, 2006) (Singh, 2005). In Singh's PhD thesis, he proposed EM-ONE, an architecture for commonsense thinking, to be capable of reflective reasoning about situations involving physical, social, and mental dimensions. EM-ONE uses as its knowledge base a library of commonsense narratives, each describing the physical, social, and mental activities that occur during an interaction between several actors. EM-ONE reasons with these narratives by applying "mental critics", procedures that debug problems that exist in the outside world or within EM-ONE itself.

A-Brain, B-Brain and C-Brain. Figure 9.1 shows Minsky's notion of multiple brain machinery. A-Brain receives signals from the external world and reacts to these signals, resulting in muscle movement. By itself, the A-Brain only reacts to external events but with no sense of what they might mean. The B-Brain is connected to A-Brain. B-Brain can affect the external world by controlling how A-Brain will react. B-Brain would need appropriate ways to represent things. C-Brain will supervise B-Brain. C-Brain acts like a manager and can give general guidance.

Six-layer structure. Minsky also proposed that the human mental resources are organized into six broad layers of processes as illustrated below Fig. 9.2.

Three of the layers are directly related to thinking. The lower two layers are more of reactive responses. And at the higher level,

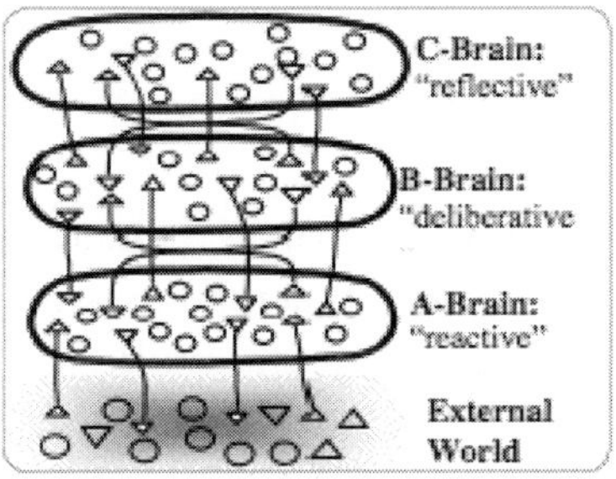

Fig. 9.1. A-Brain, B-Brain, and C-Brain.

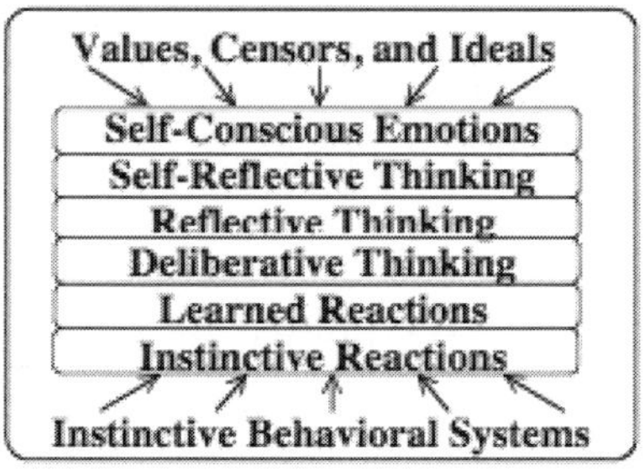

Fig. 9.2. Six-layer structure of the brain process.

the emotion is interacting with the bottom layers. Minsky explained that we are born with the instincts that help us to survive (instinctive reactions). And as we progress in life, we learn that certain conditions demand specific ways to react (learned reactions). In situations where we have to consider several alternatives and try to decide which would be best, we turn on our deliberative thinking process. When we reflect on our decision, we are not reacting to external events but what is happening inside our brain (reflective thinking). As we think about our plans to take that decision, we are engaging in self-reflective thinking. In the last stage of self-conscious emotions, Minsky explained that we would often ask ourselves what other people think of our decision and plans. In other words, do we hold up to the values that we set or meet our ideals? Then we may find conflict between how we behave and the values of those to whom we are attached, and this could lead to the kinds of cascades that Minsky called "self-conscious emotions".

Multiple Intelligences

Professor Howard Gardner, Harvard Graduate School of Education, coined the term multiple intelligences from the study of human cognition through discoveries in the biological and behavioral sciences. Gardner proposed that there are eight distinct intelligences in human, namely, linguistic intelligence, logical-mathematical intelligence, musical intelligence, bodily-kinesthetic intelligence, spatial intelligence, interpersonal intelligence, intrapersonal intelligence, and naturalist intelligence (Gardner, 1999, Gardner, 2003).

- Linguistic intelligence involves the sensitivity to spoken and written language, the ability to learn languages, and the capacity to use language to accomplish certain goals. Lawyers, speakers, writers, and poets are among the people with high linguistic intelligence.
- Logical-mathematical intelligence involves the capacity to analyze problems logically, carry out mathematical operations, and investigate issues scientifically. Mathematicians, logicians, and scientists exploit logical-mathematical intelligence.
- Musical intelligence entails skill in the performance, composition, and appreciation of musical patterns. Gardner indicated that musical intelligence is almost parallel structurally to linguistic intelligence.
- Bodily-kinesthetic intelligence entails the potential of using one's whole body or parts of the body to solve problems or fashion products. Obviously dancers, actors, and athletes foreground bodily-kinesthetic intelligence. However, this form of intelligence is also important for craftspersons, surgeons, bench-top scientists, mechanics, and many other technically oriented professionals.
- Spatial intelligence features the potential to recognize and manipulate the patterns of wide spaces as well as the patterns of more confined areas. Those who possess such intelligence include pilots, navigators, sculptors, surgeons, chess players, and architects.
- Interpersonal intelligence denotes a person's capacity to understand the intentions, motivations, and desires of other people, and consequently, to work effectively with others. Salespersons, teachers, clinicians, religious leaders, and political leaders all need acute interpersonal intelligence.

- Intrapersonal intelligence involves the capacity to understand oneself, to have an effective working model of oneself — including one's own desires, fears and capacities — and to use such information effectively in regulating one's own life.
- Naturalistic intelligence involves expertise in the recognition and classification of the numerous species in his or her environment. Those who have such intelligence are most skillful in applying the accepted "folk taxonomies" in cultures with a scientific orientation, and then to have an extensive knowledge of the living world.

Linguistics and logical-mathematical intelligence tend to dominate most tests of intelligence. This type of test may not isolate people who have other intelligences such as artistic abilities. Musical intelligence, bodily-kinesthetic intelligence, and spatial intelligence are particularly notable in the arts.

Triarchic Theory of Intelligence

This theory was proposed by Professor Robert J. Sternberg, a psychologist and psychometrician in Tufts Unversity. Sternberg believes there are three types of human intelligence (Sternberg, 2003):

- *Analytic intelligence.* This involves the ability to complete academic, problem-solving tasks such as those used in traditional intelligence tests. These types of tasks usually present well-defined problems that have only a single correct answer. People with such analytic intelligence are said to perform well in IQ tests, GAT tests, or any similar tests that only test their analytic skill that is learnt in school or through books. According to Sternberg, people with this intelligence are school smart or book smart.
- *Creative intelligence.* The ability to successfully deal with new and usual situations by drawing on existing knowledge and

skills. For example, if you are in a situation where you cannot find a proper opener to open a wine bottle, and you start to use your key attached with a hook that acts as a handle and across-bar (some other ways), to open the wine bottle, then you have used creative intelligence. Hence, unlike a task in analytic intelligence, which in Sternberg's word involves single correct answers, tasks requiring creative intelligence have open-ended or many possible answers.

- *Practical intelligence.* This involves the ability to adapt to everyday life by drawing on existing knowledge and skills. Practical intelligence is involved when dealing with everyday personal or practical problems. It may also be involved when dealing with new and unusual situations in everyday life, e.g. if you find yourself alone in an unfamiliar suburb, without money or a mobile phone, and have missed the last train or bus back to your home. According to Sternberg, successfully dealing with such a situation involves a distinctly different part of intelligence, often observed in people who are "street smart". "Street smart people can usually make these adjustments, applying their knowledge and skills in effective ways. According to Sternberg, the three parts of intelligence involve abilities that are different, separate, and are not "fixed", that is, they can change (become stronger or weaker) through experience in everyday life.

Robert J. Sternberg indicates that a person may be stronger in one or more of these three intelligence that make them successful in the society and culture they live in (where successful is according to their own individual definition of success).

When a person is sufficiently strong in each of the three parts, then the three parts will be "in balance".

Sternberg added that individuals with successful intelligence often have a "can-do" attitude, are able to learn from past experiences, and can apply their mental abilities to achieve their goals and ambitions in real-life situations. (Source http://en.wikipedia.org/wiki/Robert_Sternberg)

Conceptual Blending Theory

The conceptual blending (CB) theory was proposed by Gilles Fauconnier and Mark Turner. The theory development began in 1993, and subsequently, Fauconnier and Turner published a book entitled "The Way We Think" in 2002 that further explained the conceptual blending theory (Fauconnier and Turner, 2002). The key process in the theory is blending, i.e. humans are unconsciously but constantly blending information when talking, listening, imagining, and thinking. Blending is said to be the basic cognitive operation. CB theory claims to be related to cognitive architecture theories such as Soar, ACT-R, and frame-based theories of Marvin Minsky. The word "blending" is also sometimes used interchangeably with the word "integration," and hence, conceptual blending is also known as conceptual integration. Note that blending and integration are a kind of fusion process.

Blending is a set of mental operations for combining cognitive models in a network of discrete mental spaces. Mental spaces are said to interconnect in working memory, which can be modified dynamically, and they can be used to model dynamic mappings in thought and language. Building a conceptual blending network involves setting up several mental spaces. The following diagram shows a basic conceptual blending network.

Figure 9.3 shows the basic conceptual blending networks. The network comprises of three key spaces, namely, the input space, generic space, and the blend space. These spaces are known as mental spaces. Therefore,

a. Input mental spaces (the figure shows two input mental spaces, i.e. Input 1 and Input 2).
b. A generic mental space that captures the structure that the two input mental spaces share.
c. The blended space forms a new mental space.

Mental spaces are small conceptual packets constructed as we think and talk, for purposes of local understanding and action. Mental spaces are said to be partial assemblies containing elements, and

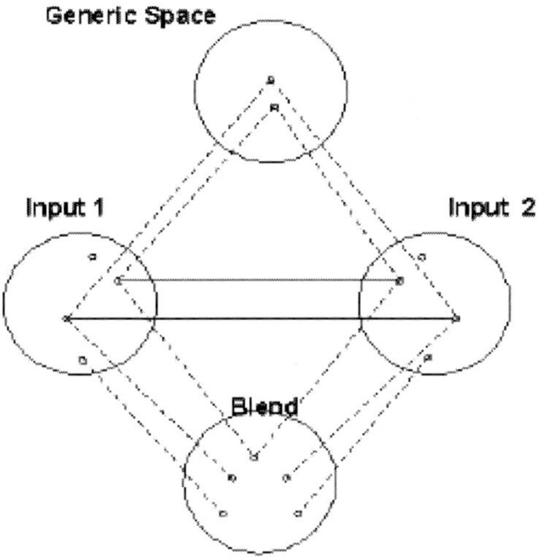

Fig. 9.3. Conceptual blending networks.

structured by frames and cognitive models. They can be used generally to model dynamical mappings in thought and language. The mental space is a theoretic construct corresponding to possible worlds in philosophy. The main difference between a mental space and a possible world is that a mental space does not contain a faithful representation of reality but an idealized cognitive model (Fauconnier and Turner, 1998, 2002).

The critical thing in CB theory is how to find the relation between the two input mental spaces that lead to the new blend. The theory calls these the "vital relations".

CB theory was first formulated for cognitive semantics. Cognitive semantics is part of cognitive linguistics, i.e. understanding of language creation, learning and usage as explained by reference to human cognition in general. And the theory also claims that this may explain the way we think.

Commonsense Knowledge Representation

The ability to think and behave intelligently requires both large quantities of common sense knowledge and the ability to learn from experience. The brain has a huge amount of internal knowledge stored, and more so, commonsense knowledge. How could a machine be able to have such commonsense knowledge? Till today, no biologically inspired techniques have been very successful in representing large-scale knowledge in a machine. Hence, a number of attempts have been made using traditional approaches to have large-scale commonsense knowledge. Three of these are discussed here, namely, Cyc, WorldNet, and Open Mind (Open Mind Common Sense).

1. Wordnet was started in 1985 in Princeton University under the direction of psychologist, Professor George A. Miller. WordNet is a semantic lexicon for the English language. It groups English words into sets of synonyms called synsets, provides short, general definitions, and records the various semantic relations between these synonym sets. The purpose is twofold: to produce a combination of dictionary and thesaurus that is more intuitively usable, and to support automatic text analysis and artificial intelligence applications. As of 2006, the database contains about 150 000 words organized in over 115 000 synsets for a total of 207 000 word-sense pairs; and in compressed form, it is about 12 megabytes in size (from http://en.wikipedia.org/wiki/Wordnet).
2. Cyc was started in 1984 by Doug Lenat. The name "Cyc" comes from "encyclopedia". The Cyc project is by far the largest and most ambitious attempt to build a database of commonsense knowledge, and today, Cyc contains over two million facts and rules about the everyday world. Its ontology is the most expressive presently available, and represents the state-of-the-art in broad coverage of formal knowledge representation. It includes a wide range of ideas about representing such matters as space, time, beliefs, goals, social relationships, physical constraints, as well as many other domains. (From http://en.wikipedia.org/wiki/Cyc.)
3. Open Mind (Open Mind Common Sense), started in 1999, is based at the Massachusetts Institute of Technology Media Lab. The goal is to build a large common sense knowledge base

> from the contributions of many thousands of people across the Web. Since its founding in 1999, it has accumulated more than 700000 English facts from over 15000 contributors. The knowledge collected by Open Mind Common Sense has enabled dozens of research projects at MIT and elsewhere. The ConceptNet and LifeNet commonsense knowledge bases are derived from parsing the collected data. (From http://en.wikipedia.org/wiki/Open_Mind_Common_Sense.)

"Other similar quests, such as MindPixel, General Purpose Solver (GPS), EM-ONE and Knowledge Machine are briefly presented in the Notes for Chapter 9."

WordNet, Cyc, and OpenMind all have the same goal of trying to achieve commonsense knowledge. There, each tries to assemble a comprehensive ontology and database, and to enable AI applications to perform human-like reasoning using the rich language captured. They each has slight variations. For example, WordNet only has some common sense knowledge while Open Mind Common Sense has a large common sense knowledge base as this is MIT media lab's main goal (MIT media lab maintains and develops the Open Mind Common Sense). The WordNet is a semantic lexicon for the English language and its chief purpose is to produce a combination of dictionary and thesaurus and to support automatic text analysis and artificial intelligence applications.

The three common problems WordNet, Cyc, and OpenMind need to overcome are:

a. Handling of exceptions. How to handle exceptions? For example, commonsense knowledge by default rule may indicate that "All birds can fly". But penguins and ostriches are examples of birds that cannot fly. How to enable the commonsense knowledge to handle such exceptions? Humans have the capability of handling exceptions. Humans can also learn by just watching.
b. Commonsense logic. How to build analog or equivalent commonsense functions? For example, putting a cardboard on top

of the head instead of carrying an umbrella, could be a temporary solution during a light drizzle.

c. Robust commonsense database. How to build robust commonsense data inside a computer system? This includes how to have a good design to ensure continuous encapsulation of the knowledge efficiently and effectively without incurring scalability problem.

Currently, we have not been successful in finding ways to teach computers all this commonsense knowledge. It is said that human as young as five to seven years old already have much common sense knowledge.

The Wise Machine — AI with Wisdom

The ultimate desire of thinking is to achieve "wisdom". Wisdom is the ability to make sound judgments and good decisions, foresee consequences and draw important lessons from situations. Now, can machines have wisdom? Can a machine read a book such as Moby-Dick and draw important lessons, such as "One should not be too focused and obsessed with one goal to the point that one excludes the more important things in life"? (Note: Moby-Dick was written by Herman Melville in the eighteenth century.) Or can a machine be so creative as to devise the next solution to solve the increasing bottleneck database issue, such as a cache oblivious algorithm? Note: a cache oblivious algorithm is designed to exploit the CPU cache without having the size of the cache (or length of the cache lines) as an explicit parameter.

I have searched various publications and not come across many materials that discuss this topic and also with respect to building intelligent systems. Perhaps "wise" or "wisdom", the word itself, is not well understood. What is wisdom and what does it mean to be a wise man?

In the Bible, three books (Job, Proverb, and Ecclesiastes) touch on wisdom. From studying these books, one can draw a conclusion that wisdom is a principle and it can be further broken down into 60 desirable character traits. Character is wisdom, but wisdom is not character. Why is this so?

> "*Get wisdom, get understanding; do not forget my words or swerve from them. Do not forsake wisdom, and she will protect you; love her, and she will watch over you. Wisdom is supreme; therefore get wisdom. Though it cost all you have, get understanding.*" — Proverb 4:5–7.

> "*Wisdom is the principal thing; therefore get wisdom.*" (Proverb 4:7)

The wisest man, according to the Bible, who ever lived is King Solomon.

1 King 3:16–28. An example of wisdom from the Bible on King Solomon's wise decision. This story goes as follows:

Now, two prostitutes came to the King Solomon and stood before him. One of them said, "My lord, this woman and I live in the same house. I had a baby while she was there with me. The third day after my child was born; this woman also had a baby. We were alone; there was no one in the house but two of us.

During the night, this woman's son died because she lay on him. So she got up in the middle of the night and took my son from my side while I, your servant, was asleep. She put him by her breast and put her dead son by my breast. The next morning, I got up to nurse my son — and he was dead! But when I looked at him closely in the morning light, I saw that it wasn't the son I had borne."

The other woman said, "No! The living one is my son; the dead one is yours."

But the first one insisted, "No! The dead one is yours; the living one is mine." And so they argued before King Solomon.

The king said, "This one says, "My son is alive and your son is dead," while that one says, "No! Your son is dead and mine is alive."

Then the king said, "Bring me a sword." So they brought a sword for the king. He then gave an order: "Cut the living child in two and give half to one and half to the other."

The woman whose son was alive was filled with compassion for her son and said to the king, "Please, my lord, give her the living baby! Don't kill him!"

But the other said, "Neither I nor you shall have him. Cut him in two!"

Then the king gave his ruling: "Give the living baby to the first woman. Do not kill him; she is his mother."

When all Israel heard the verdict the king had given, they held the king in awe, because they saw that he had the wisdom from God to administer justice.

This short story illustrates King Solomon's wisdom in decision making (the ability to make good decisions, weigh options, and act prudently). It also shows the human emotion in communication and how Solomon wisely tapped on human emotions to make his wise decision.

> "*To understand wisdom fully and correctly probably requires more wisdom than any of us have*," Robert Sternberg said.

Pattern recognition and wisdom. The work by Nobel laureate Herbert Simon and others has shown that pattern recognition is among the most powerful, and perhaps the foremost mechanism of successful problem solving. The brain machinery appears to have hardwired some of this ability early in life, and as life progresses, the external environment sharpens or shapes this pattern-recognition process and enables it to be fully operational (as discussed in the chapter on gene versus environment).

It is said that the brain comes pre-wired for certain kinds of pattern recognition but not others. The visual cortex, sematosensory cortex, auditory cortex, and motor cortex, to a large degree, are pre-wired. The more complex cortical regions such as the association cortex may have relatively less pre-wired knowledge. It can possess a huge capacity to process any kind of information. This is the part where the mind is open (with an **open-ended-open-minded design)** for new input beyond the pre-wired.

Could a certain kind of **pattern-recognition mechanism** in our brain capture the "wisdom" that we do not yet know?

Knowledge, character, and wisdom.

> "*Many people know a great deal but are all the more foolish because of it. They have not yet learned how to apply the knowledge they have. For the successful conduct of life, mere knowledge is not enough.*" — Anonymous.

Knowledge is frequently equal to power (common saying: knowledge is power; knowledge can bring us to the technological edge, etc), but knowledge alone does not often bring a person to success. Most studies have shown what make people successful are most part, character. Here are 60 ideal character traits any parent would desire their children to have. And I assume any scientists, likewise, would like their intelligent robots to ultimately be able to model after some of these character traits. These are, to be: 1. Appreciative; 2. Attentive; 3. Available; 4. Committed; 5. Compassionate; 6. Concerned; 7. Confident; 8. Considerate; 9. Consistent; 10. Content; 11. Cooperative; 12. Courageous; 13. Creative; 14. Decisive; 15. Deferent; 16. Dependable; 17. Determined; 18. Diligent; 19. Discerning; 20. Discreet; 21. Efficient; 22. Equitable; 23. Fair; 24. Faithful; 25. Fearless; 26. Flexible; 27. Forgiving; 28. Friendly; 29. Generous; 30. Gentle; 31. Honest; 32. Humble; 33. Joyful; 34. Kind; 35. Loyal; 36. Meek; 37. Merciful; 38. Observant; 39. Optimistic; 40. Patient; 41. Peaceful; 42. Perseverant; 43. Persuasive; 44. Prudent; 45. Punctual; 46. Purposeful; 47. Resourceful; 48. Respectful; 49. Responsible; 50. Secure; 51. Self-controlled; 52. Sincere; 53. Submissive; 54. Tactful; 55. Temperate; 56. Thorough; 57. Thrifty; 58. Tolerant; 59. Truthful; 60. Virtuous.

> "*Knowledge comes by taking things apart: analysis. But wisdom comes by putting things together.*" — John A. Morrison. Wisdom is the proper use of knowledge. How can we design a machine to use

its knowledge to derive wisdom? How can we design a machine that has human-like character traits?

"*The rod of correction imparts wisdom, but a child left to himself disgraces his mother.*" — Proverb 29:15.

Machine with vision. The Bible says, "*Where there is no vision, the people perish.*" — Proverbs 29:18. And a Japanese proverb says, "Vision *without action is daydream. Action without vision is nightmare.*" What does it mean to have a vision or an intelligent foresight? Here, I am not referring to the faculty of sight but the ability to discern things or have an unusual competence to perceive and set goals, objectives, and plan for the future.

How to create a faculty that could have such intelligent ability? In what situation will we produce great vision? Does it happen more often during the time of suffering or pain? Is the ability to have vision related to emotion? There seems to be some connection to emotion but we are not sure how it works.

Summary

To create a machine with true thinking ability like a human is currently unimaginable. However, if one could truly make a "program" that changes, adapts, or has "plastic" abilities, and it can alter its output behavior due to the "program plasticity", we may one day say that the machine has some "thinking" capability and "a mind of its own".

"Thinking" is a pretty high level concept; this chapter discusses some of the high level theories of the brain's machinery proposed by scientists. The high level theories may potentially explain how we think and how we can make a machine "think" like a human. These include:

- Multiple layers of machinery
- Theory of multiple intelligences

- Triarchic theory of intelligence
- Conceptual blending theory

To think, one needs to have much commonsense knowledge and to constantly build up this knowledge, in addition to intelligence abilities. Cyc, WorldNet, and OpenMind are some examples of the human attempt to build large-scale commonsense knowledge.

Lastly, the purpose of thinking is to achieve wisdom. Wisdom is a principal thing, and hence, it is related to character. Could we have a machine with character and wisdom? Currently, automatic character building based on environmental influence looks difficulty to achieve and we do not know of any research carried out in this area.

Chapter 10

MODELING THE ENTIRE BRAIN: BIOLOGICALLY INSPIRED COGNITIVE ARCHITECTURES

To model the entire brain or engage in any large scale modeling of the brain is not a trivial task.

Figure 10.1 shows one possible way of categorizing to enable us to discuss the level of detail in modeling of the brain. It is divided into six levels and further subdivided into three bands.

At the low bands, the modeling is most complicated. I do not know of any attempt to model at the atomic and molecular levels such as molecular[a] interaction and protein folding for the whole brain.

At the medium band are the cellular and network levels. This band is also relatively complicated to model but some modeling techniques are suggested, as discussed in the earlier chapter.

The last band is at the system and organismal levels. Modelings at this band typically consider the functional, cognition, behavior, and the affect aspect. In some sense, it is a high level cognitive process. At the system level, different regions of the brain can be modeled to interact with other regions and work together to give cognition. The system level can take into account properties such

[a] A molecule is group of atoms chemically bonded together to form a unit.

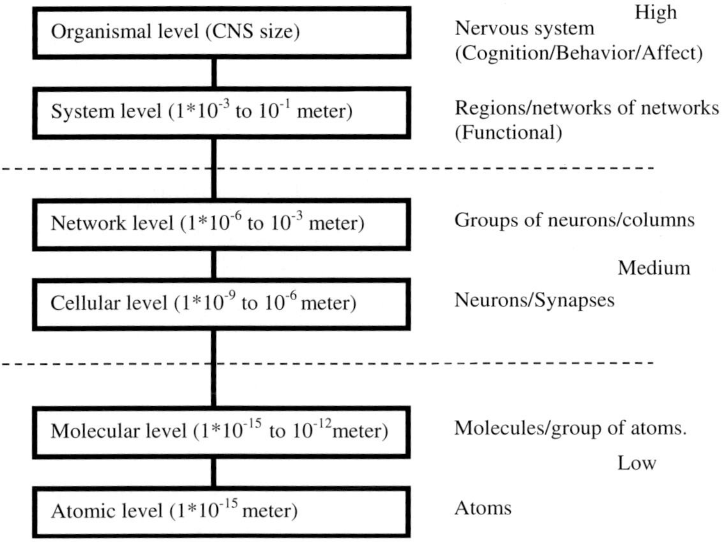

Fig. 10.1. Levels of modeling the brain.

as the complementary role, top-down expectation, bottom-up influence, etc. At the organismal level, one attempts to study and understand the whole system output and model after the system's cognition, behavior, and affect. These models typically consider long-term memory, working memory, the ability to perform reasoning, planning, goal-setting, and learning.

Blue Brain project by IBM and CCortex project by Artificial Development are two of the examples from commercial companies attempting to model the entire brain at the cellular level. Many other companies are involved at different levels of modeling the brain to build intelligent systems. A few such companies in United States include eCortex Inc (http://e-cortex.com/), Numenta Inc (http://www.numenta.com/), Cyberkinetics/Neurotechnology System Inc (http://www.cyberkineticsinc.com/content/index.jsp), and Charles River Analytics (http://www.cra.com/about-us/index.asp).

The Cognitive Computing group at IBM Almaden Research Center claims to have completed simulating the mouse brain on the

BlueGene/L supercomputer (BBC News, April 27, 2007). They simulated the mouse cortex with a massive parallel cortical simulator that incorporates relatively simple single compartment spiking neurons, spike-timing dependent plasticity (STDP) and axonal delays (Ananthanarayanan and Modha, 2007). With the completion of the Blue Brain project, IBM on 20 Nov 2008, announced funding from DARPA to build "cognitive computing" with its collaborators under a grant of $4.9 m. (See http://www-03.ibm.com/press/us/en/pressrelease/26123.wss for more information.)

Henry Markram, in his report on the Blue Brain project, indicated that to model the brain at the atomic level (atomic collision is simulated or performed at the atomic scale) would take days to simulate a microsecond of protein folding in the most powerful supercomputer. However, models at higher levels such as the molecular or cellular levels can capture lower-level processes and allow complex large-scale simulations of biological processes (Markram, 2006).

Note that the mouse brain has 100 million neurons and the Blue Brain project simulated at the cellular level is doing that. The human brain has about 23 billion neurons (Rabinowicz *et al.*, 2002). To simulate that, we would need today's computer to increase its computational power by about one million fold before we are able to simulate human brain at the cellular level.

Modeling at the atomic and molecular levels is currently very difficult, because, in the brain, every molecule is performing complex tasks that need a powerful computer to simulate. Besides that, the simulation and modeling process would also need to capture all the functions over trillions and trillions of these molecules as well as all the rules that govern how they interact.

The next closest biological form of modeling at the cellular level is the Hodgkin–Huxley's original cell membrane equation, a four-dimensional equation. However, the Hodgkin–Huxley's cell membrane equation is based on a giant squid axon. In the human brain, there are billions of neurons and it is impossible at our current technology to measure every single neuron to find out all the parameters and values to form every cell membrane question, taking into consideration their ion channels, different types of synapses, the specific spatial geometry of individual neurons, etc. It is equally impossible to compute them at the current computing power.

Hence, modeling at the cellular level using the Hodgkin–Huxley type neuronal models on the human brain is equally complex. A simplified form of cellular level modeling is then proposed, such as the two-dimensional neuron models, which is a reduction form of the four-dimensional Hodgkin and Huxley's cell membrane equation. Other simplified neuron models that are more commonly used are the spiking neuron models such as the integrate-and-fire model and spike response model (Gerstner and Kistler, 2002). Grossberg's shunting networks equation is another network variant form of the neuron model following Hodgkin and Huxley's cell membrane equation. The key challenges in all these simplified models are how biologically plausible the models are with respect to the anatomic data of the cell membrane equations for the specific region in the brain, the parameters, and the parameters' values selection and setting. Even then, scaling the model itself up to the brain level is complex.

Hence, to model the entire brain, the most current approach adapts the higher cognitive level, i.e. at the system or cognition level. In this chapter, we will present the six different types of cognitive architecture, which serve to model the entire brain at the system and cognition level. Section 1 will give an overview of the integrated cognitive architecture approach to model the entire brain. Sections 2, 3, 4, 5, 6, and 7 will discuss the cognitive architecture of SOAR, ACT-R, ICARUS, BDI, Subsumption, and CLARION respectively (Chong *et al.*, 2007).

Integrated Cognitive Architecture

An integrated cognitive architecture can be defined as a single system that is capable of producing all aspects of behavior, while remaining constant across various domains and knowledge bases (Newell, 1990; Anderson *et al.*, 2004). This system would consist of many modules (or components) working together to produce a behavior. These modules contain representations of knowledge, memories for storage of content, and processes utilizing and acquiring knowledge. Integrated cognitive architectures are often used to explain a wide range of human behavior, and to mimic the broad capabilities of human intelligence (Anderson, 2004; Langley and Choi, 2006).

Research on integrated cognitive architectures is interdisciplinary by nature, spanning the fields of artificial intelligence, cognitive psychology, and neurobiology. Over the past decades, many cognitive architectures have been proposed and steadily developed, based on different approaches and methodologies. However, despite integrated cognitive architecture having been an important research area, there is a lack of formal reviews on the subject domain.

While it is clearly impossible for this chapter to include all existing research and individuals who have contributed significantly to the topic, we aim to cover a good mix of the systems that are representative of various approaches and disciplines.

Figure 10.2 shows an overview of the six cognitive architectures, namely, Soar, ACT-R, ICARUS, BDI, Subsumption, and CLARION, roughly classified according to their roots and emphasis.

Soar, based on the physical symbolic hypothesis (Newell, 1990), is one of the earliest and most extensively developed AI architectures in history. ACT-R (Anderson *et al.*, 2004) and ICARUS (Langley and Choi, 2006), on the other hand, were developed with the aim of

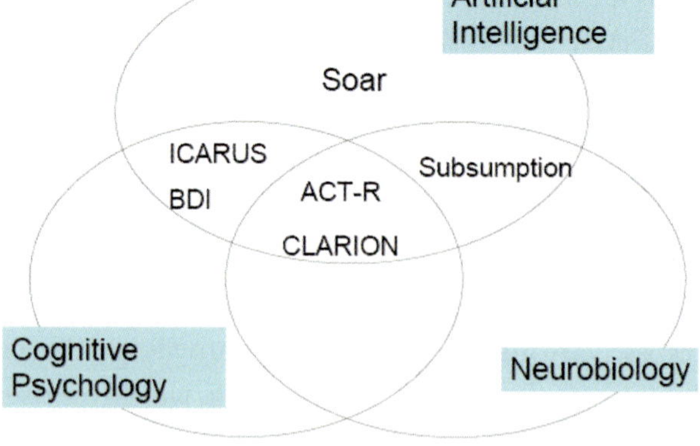

Fig. 10.2. Overview of the six cognitive architectures.

producing artificial intelligence mimicking human cognition. While the three architectures share many features of classical artificial intelligence, like symbolic representation, production rule-based inference, and means-end analysis for problem solving, ACT-R, and ICARUS are notably different from Soar by their strong emphasis of producing a psychologically motivated cognitive model.

BDI architecture is a popularly used framework incorporating beliefs, desires, and intentions for designing intelligent autonomous agents (Bratman *et al.*, 1988; Rao and Georgeff, 1991). Based on the studies of folk psychology and intentional systems, BDI has a special focus on intentions, representing an agent's commitments to carry out certain plans of actions (Georgeff and Ingrand, 1989).

Similar to the above, CLARION cognitive architecture follows the traditional approach in computer science. However, CLARION is a hybrid model integrating both symbolic and connectionist processing of information (Sun and Peterson, 1996; Sun and Zhang, 2006).

Furthermore, CLARION is based on a neural network in the design of the architecture as well as cognitive psychology. As a result, it is similar to ACT-R cognitive architecture as both models are based on the combination of artificial intelligence, cognitive psychology, and neurobiology.

Coined as the new artificial intelligence, the Subsumption architecture is drastically different from other cognitive architectures in its approach. Subsumption architecture is behavior-based and thus does not contain any problem solving or learning modules. The idea of higher layers subsuming lower layers in Subsumption architecture also originate from neurobiology, therefore both artificial intelligence and neurobiology form the basis for Subsumption architecture (Brooks, 1999; Toal *et al.*, 1996).

SOAR Architecture

SOAR stands for State, Operator, and Result. Soar, one of the first cognitive architecture proposed, is a general cognitive architecture for developing systems that exhibit intelligent behavior and it has been under continuous development since the early 1980s (Laird *et al.*, 1987). Soar has generated a wide spectrum of applications particularly in military operations such as virtual training for large scale combat flight simulation to generate intelligent human-like behavior when viewed by a training audience participating in operational military exercises (Jones *et al.*, 1999) and urban combat training (Wray *et al.*, 2004). The current working prototype Soar system is in version 8.6 (Laird *et al.*, 2006).

Figure 10.3 illustrates the Soar architecture (Lehman *et al.*, 2006) consisting of memory structures and a decision-making mechanism linking perception to action. The memory structures present in the Soar architecture include long-term memory and working memory. Information from the environment is made available in the working memory via perception, allowing appropriate actions to be chosen during domain-independent problem solving. The external environment can also be influenced by the architecture through the implementation of selected actions.

Knowledge is stored in the long-term memory, which can be classified into procedural, semantic, and episodic memories. Procedural memory provides the knowledge of performing tasks.

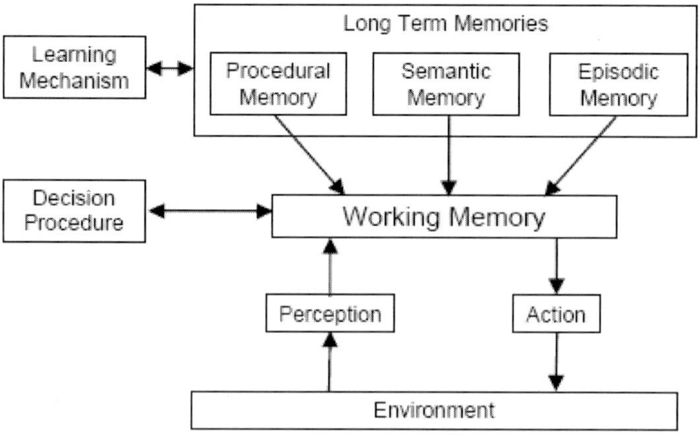

Fig. 10.3. The Soar cognitive architecture.

Semantic memory stores general facts about the world (e.g. a car has four wheels and is encoded as declarative structures) and is considered declarative knowledge. On the other hand, episodic memory contains specific memories of an event experienced. Hence, both procedural and semantic memories are universally applicable, whereas episodic memory is contextual specific. In instances whereby procedural knowledge is insufficient, semantic and episodic memories are employed as cues to aid in problem solving.

The working memory in Soar's cognitive architecture houses all the knowledge that is relevant to the current situation. It contains the goals, perceptions, hierarchy of states, and operators. The states (and sub-states) give the information on the current situation. The operator provides the steps to apply during problem solving, while the goal directs the architecture into the desired state. Contents of the working memory, also known as working memory elements, can trigger both the retrieval of relevant knowledge from the long-term memory into the working memory, and motor actions.

Soar's functions and processes. The Soar agent makes use of means-ends analysis for problem solving, which requires the system to select and apply operators to result in a new state that is closer to the desired state. In order to bring the system closer to its goal, the Soar agent implements a five-phase decision cycle, constituting of input, elaboration, decision procedure, application, and output. The main function of the decision cycle is to choose the next operator to apply (Laird *et al.*, 1986a; Lehman *et al.*, 2006).

Percepts are added to the working memory in the input phase for use during the elaboration phase. Production rules are then matched with the working memory elements in order to bring knowledge relevant to the current problem into the working memory. Meanwhile, preferences are created to act as recommendations for the selection of appropriate operators. The elaboration phase continues till the firing of rules in the long-term memory ceases, ensuring that all knowledge relevant to the current situation is considered before a decision is made (Laird *et al.*, 1986b; Lehman *et al.*, 2006). When the elaboration phase reaches quiescence, the decision cycle proceeds to the decision procedure where preferences are evaluated. The best suggested operators are chosen and applied during the application phase. A motor action is then performed during the output phase as a result of applying the selected operator.

An impasse is encountered whenever procedural knowledge is inadequate for problem solving (Laird *et al.*, 1986b). In Soar, there are four types of impasse: no-change, tie, conflict, and rejection. A no-change impasse occurs when the elaboration phase enters quiescence without any suggestion, while a tie impasse refers to the situation whereby no object is superior over the others. Cases in which two or more candidate objects are better than each other result in a conflict impasse. A rejection impasse is encountered when all operators are rejected.

Any impasse encountered provides an opportunity for the Soar agent to learn from its experience. The learning mechanisms proposed in the Soar architecture are chunking, reinforcement learning, episodic memory, and semantic memory. Successful resolution of the impasse results in the termination of the goal or sub-goal, which in turn leads to the formation of chunks. The chunking mechanism enables new production rules to be added to the long-term memory. These chunks are used whenever a similar situation is encountered, thereby avoiding the same impasse and improving the performance of the agent in the future.

The Soar agent also receives rewards from success or punishments from failure, allowing the agent to undergo reinforcement learning. Any operator resulting in a reward upon execution is given positive reinforcement, and such operators are more likely to be selected in the future. Episodic and semantic memories store information on the agent's past experiences, and therefore, both are used as additional cues to select applicable operators (Laird *et al.*, 1986b; Lehman *et al.*, 2006).

ACT-R Architecture

ACT-R stands for Adaptive Control of Thought-Rational. ACT-R is pioneered by Professor John Anderson from Carnegie Mellon University, Psychology Department in early 1990s (Anderson, 1993).

ACT-R was developed as a model of human cognition using empirical data derived from experiments in cognitive psychology and brain imaging. The ACT-R design concept assumes that cognition of the brain emerges through the interaction of a number of independent modules. Some examples of the independent modules are the visual module, control module, memory module, and motor module.

In order for the modules to communicate, a common knowledge representation is needed. In ACT-R, knowledge is represented by chunks, a simple symbolic representation system. Each chunk has a number of slots, each of which contains a single symbol. These symbols can represent anything (including other chunks), but do not have an inherent semantic value.

ACT-R provides a step-by-step simulation of human behavior for detailed understanding of human cognition. It consists of four basic modules, as well as a central production system (see Fig. 10.4). They are, namely, the perceptual, motor, intentional, and memory

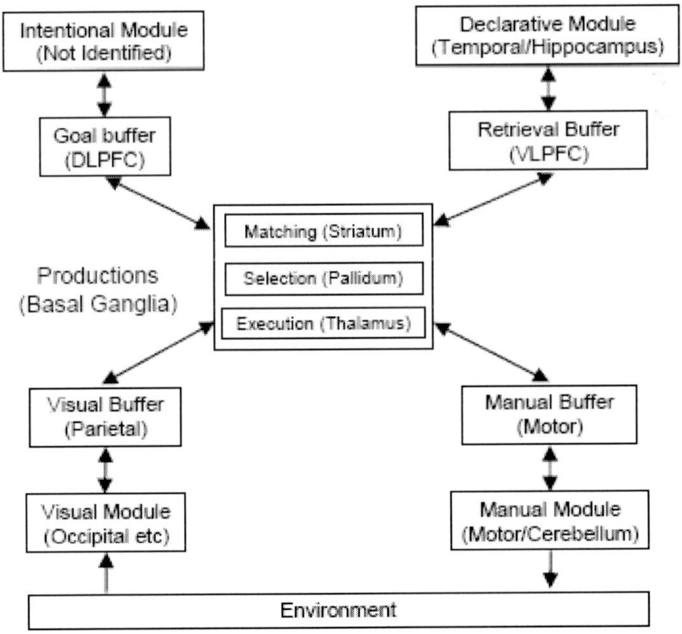

Fig. 10.4. The ACT-R cognitive architecture (Anderson *et al.*, 2004).

modules. The perceptual module receives sensory input from the external world and the motor module executes the actions produced by the production system. The intentional module maintains the task-related objectives during the cognition process and the memory module contains the long-term memory. Each module has a buffer to keep track of one's internal state during problem solving. The central production system generates production rules based on the information available in the buffers of the four basic modules (Anderson *et al.*, 2004).

Each module of the ACT-R architecture has its plausibility in neurobiology. They are summarized as follows:

- *Visual module.* The function of the visual module is to receive the visual input from the external environment. The obtained information is used for processing in the production system. In neurobiology, the occipital lobe of the forebrain is responsible for the vision information processing.
- *Motor module.* The motor module is for motoring actions, taken in the production system. The counterparts in the neurobiology are the motor cortex and cerebellum. The motor cortex is the region of the cerebral cortex involved in the planning, control, and execution of voluntary motor functions, and the cerebellum controls the body movement in space and coordinates all motor activities and postures.
- *Intentional module.* The intentional module consists of the context, goals, and schedules in problem solving. In the human brain, the posterior parietal cortex plays a major role in maintaining the problem state. Besides, the prefrontal cortex is involved in working memory, problem solving, initiation, judgment, impulse control, and social and sexual behaviors. It is thought to be an important part in the maintaining of the problem context.
- *Memory module.* The memory can be categorized into sensory, short-term (or working memory), and long-term memories according to the effective time span that memory can be recalled. In terms of the nature of information stored, the long-term

memory can be divided into declarative (explicit) and non-declarative (implicit) memories. In the human brain, the frontal and temporal lobes, where the hippocampus is located, are responsible for long-term memories.

- *Production system.* According to the ACT-R theory, the basal ganglia and its associated connections are thought to implement the production system. There are three parts of the basal ganglia that play important roles in the production system. They are the striatum, pallidum, and thalamus. The striatum acts as a pattern recognition function that matches the input patterns with that in the memory under the problem context. The pallidum serves as a conflict-resolution function and the thalamus is used to control the execution of the selection actions.

ICARUS Architecture

The ICARUS is a relatively new cognitive architecture proposed by Langley and Choi. The ICARUS cognitive architecture focuses on physical and embodied agents, integrating perception and action with cognition. This cognitive architecture also aims to unify reactive execution with problem solving, combine symbolic structures with numeric utilities, and learn structures and utilities in a cumulative manner. ICARUS shares a number of central features as other cognitive architectures (Langley and Choi, 2006).

The design for ICARUS has been guided by the following principles, which enable the differentiation of ICARUS from other cognitive architectures: (1) Cognitive reality of physical objects; (2) Cognitive separation of categories and skills; (3) Primacy of categorization and skill execution; (4) Hierarchical organization of long-term memory; (5) Correspondence of long-term or short-term structure; and (6) Modulation of symbolic structures with utility functions.

Architecture. Figure 10.5 presents an overview of the ICARUS architecture. The main components of the ICARUS architecture are the perceptual buffer, conceptual memory, skill memory, and motor

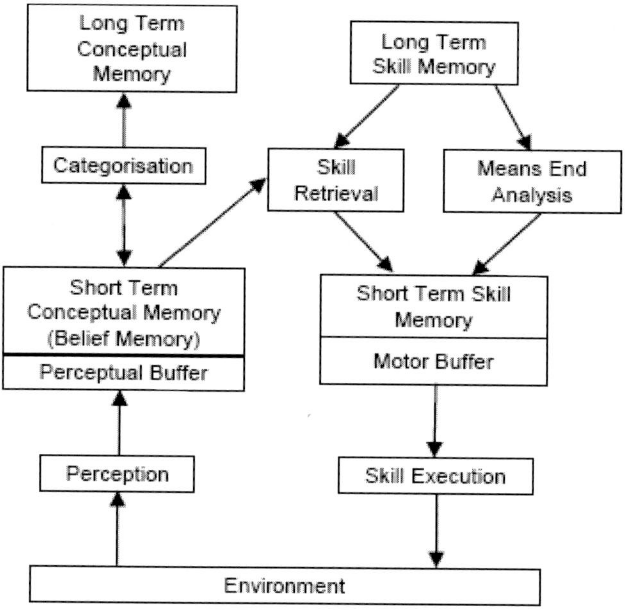

Fig. 10.5. The schematic diagram of the ICARUS cognitive architecture.

buffer. The perceptual buffer is involved in the temporary storage of percepts. The conceptual memory can be further classified into conceptual short-term and conceptual long-term memories. Conceptual short-term memory, otherwise known as belief memory, contains high-level inferences describing the relations among the objects perceived. On the other hand, conceptual long-term memory consists of conceptual structures describing classes of environmental situations. Similar to conceptual memory, skill memory is subdivided into short-term and long-term skill memory. All the skills that can be executed by the ICARUS agent are stored in the long-term skill memory, and any skill chosen to be implemented is brought into the short-term skill memory. The skill signals in the motor buffer, read out from the short-term skill memory, are then executed by the ICARUS agent to bring about changes to its environment.

Functions and processes. The key processes in ICARUS include conceptual inference, goal selection, skill execution, problem solving, and learning. Conceptual inference is the mechanism responsible for matching conceptual structures against the percepts and beliefs. It is dependent on the content and representation of elements that have been stored in both the short and long-term memories. The environment, which the ICARUS agent is situated in, affects the perceptual buffer and thus the conceptual memory, leading to a repeat of the conceptual inference that updates the descriptions of the environment. In each cycle, the agent retrieves the attributes of the perceived objects into the perceptual buffer and matches them against the conceptual definitions in the long-term memory. As a result, all the elements that are implied deductively by the percepts and the conceptual definitions are added to the belief memory. Conceptual inference occurs in a bottom-up (data driven) manner, and is similar to the elaboration phase in Soar (Langley *et al.*, 2004; Langley and Choi, 2004).

Goals in ICARUS represent the objectives that an agent aims to satisfy. In order to achieve the goals, stored skills within the long-term skill memory are executed, thereby altering the environment to bring the agent closer to its goals. ICARUS focuses on only one goal at a time. Therefore, in each cycle, only the highest priority goal not yet satisfied is attended to by the ICARUS agent. The agent then selects skills that are applicable to the agent's current belief and executes the skills chosen via the motor buffer.

Whenever the agent is unable to find an applicable skill to achieve its current goal upon reaching an impasse, it employs means-ends analysis as its problem solving strategy. Means-ends analysis involves the decomposition of a problem into sub-goals. An applicable skill path is selected for each sub-goal. After completing a sub-goal, the agent returns to the parent goal before selecting another skill path for the next sub-goal.

ICARUS is similar to Soar in terms of learning, since both architectures learn from the impasse. The difference is that ICARUS retrieves information from the goal stack to select an applicable skill, while Soar uses task-dependent knowledge to solve an impasse. Skill learning in ICARUS occurs as a result of problem solving whereby a skill is learnt when the agent is able to execute the action successfully (Langley and Choi, 2004).

BDI Architecture

BDI agent is based on the theory of human practical reasoning (Bratman *et al.*, 1988) and Denett's theory of intentional systems. Originally developed as a system that can reason and plan in a dynamic environment, BDI meets real-time constraints by reducing the time used in planning and reasoning. BDI is designed to be situated, goal directed, reactive, and social. This means a BDI agent is able to react to changes and communicate in their embedded environment, as it attempts to achieve its goals. Mechanisms for responding to new situations or goals during plan formation for general problem solving and reasoning in real time processes are also included in BDI systems (Georgeff and Ingrand, 1989; Sardina *et al.*, 2006)). BDI agents are typically implemented as a Procedural Reasoning System (PRS). Later implementations of BDI architectures are also largely based on PRS (Guerra-Hernndez *et al.* 2004; Sardina *et al.*, 2006).

The database contains the beliefs or facts about the world as well as inference rules that lead to the acquisition of new beliefs; therefore, the database consists of both the beliefs and the plan library. The beliefs are updated by the perception of the environment and the execution of intentions (Georgeff and Ingrand, 1989; Guerra-Hernandez *et al.*, 2004).

Plans refer to the sequences of action a BDI agent can perform to achieve one or more of its intentions. It is a type of declarative knowledge containing ideas on how to accomplish a set of given goals or react to certain situations. A plan consists of a body and an invocation condition. The body of a plan comprises of possible courses of action and procedures to achieve a goal, while the invocation condition specifies prerequisites to be met for the plan to be executed, or to continue executing the plan. The plan can also

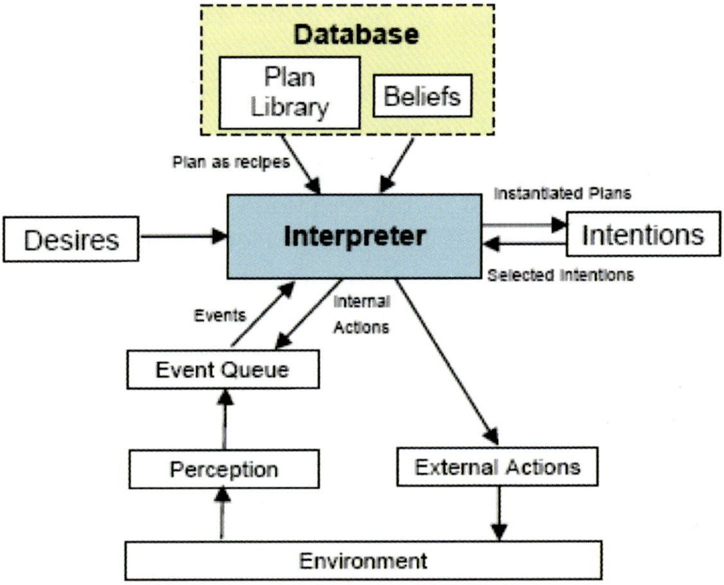

Fig. 10.6. Schematic of the BDI cognitive architecture in dMARS specification.

consist of sub-goals to achieve, and the execution of the plan may create new sub-goals, leading to the formation of goal-sub-goal hierarchy (Sardina *et al.*, 2006).

Desires, also known as goals, are objectives that a BDI agent aims to accomplish. The usage of the term "goal" in this architecture is only restricted to non-conflicting desires. Conflicts among desires are also resolved using rewards and penalties (Dastani *et al.*, 2001). A goal is said to be successfully achieved when a behavior satisfying the goal description is being executed.

Intentions are desires that the agent is committed to. Thus, it contains all the tasks chosen by the system for execution. Each intention is implemented as a stack of plan instances, and is only considered executable if the context of the plan matches with the consequence of the beliefs.

An event queue in BDI architecture refers to perceptions of the agent mapped into events stored in the queue. An event occurs when there is an acquisition or removal of belief, reception of messages, and acquisition of new goals. Therefore, events are inputs to the system.

Functions and processes. A system interpreter manipulates all components of the BDI architecture described above. It updates the event queue by perception and internal actions to reflect the events observed, followed by the selection of events. New possible desires are then generated by finding relevant plans in the plan library for the selected event. From the set of relevant plans, an executable plan is then selected and an instance plan is thus created. The instance plan is pushed onto the existing (when the plan is triggered by an internal event such as another plan) or new (when the plan is triggered by an external event) intention stack. The BDI agent interacts with its environment either through its database when new beliefs are acquired or through actions performed during the execution of the intention. The interpreter cycle repeats after the intention is executed. However, only the acquisition of a new belief or goal leads to an alteration of plans and causes the agent to work on another intention.

The BDI architecture integrates means-ends reasoning with the use of decision knowledge in its reasoning mechanism. The system performs means-ends reasoning and other planning in the context of existing intentions. A special feature of the BDI architecture is that once it is committed to a process of achieving goals, it does not consider other pathways although they may be better than the one chosen. In this way, the architecture saves decision time.

Goals in BDI architecture are dropped once the system recognizes goals as being accomplished or unable to be readily accomplished. Only when all attempts at achieving goals by trying all applicable plans have failed can the goals be labeled "cannot be readily accomplished".

The original BDI architecture lacks the mechanism of learning. However, there have been recent attempts to include learning mechanisms into BDI systems. For example, Norling (Norling, 2004) extended BDI for learning recognition primed decision making. A table lookup version of Q-learning was adopted to learn reactive rules for path finding in a grid world. More recently, Subagdja and Sonenberg (Subagdja and Sonenberg, 2005) further extended the BDI architecture to incorporate learning by generating and testing hypotheses for the purpose of formulating plans. At the multi-agent level, (Guerra-Hernandez *et al.*, 2004) expanded the BDI architecture to incorporate learning in multi-agent systems using a first order method called induction of decision tree.

Subsumption Architecture

Subsumption architecture is a "new" concept in artificial intelligence derived from behavior-based robotics. In view that classical artificial intelligence often has the problems of extensibility, robustness, and achieving multiple goals, Subsumption architecture was proposed as an incremental and bottom-up approach to deal with these problems. Subsumption architecture decomposes a problem in terms of behaviors exhibited by the robots instead of stages of information flowing within the controller as in a traditional AI design. Conversely, some researchers believed that traditional cognitive architectures would be useful for higher cognitive functions, while Subsumption architecture was only meant for reflexive tasks (Brooks, 1999; Butler, 2001; Amir and Maynard-Zhang, 2004).

Architecture. Figure 10.7 depicts the Subsumption architecture, comprising of a hierarchy of "simple" behavior-based modules organized into layers (levels of competence). Subsumption allows all layers to access the sensor's data and multiple simple behaviors to operate in parallel. Each layer is capable of controlling the system by itself, unlike classical artificial intelligent agents. Each level of competence displays a behavior to pursue a particular goal and a higher-level layer tends to subsume the underlying layers. Lower layers work like fast-adapting mechanisms, allowing the agent to react quickly to changes in its environment. In contrast, higher layers control the system towards the overall goals. As a result, lower level behaviors are the results of reaction towards the environment, while higher-level behaviors aim to pursue the primary goals. In

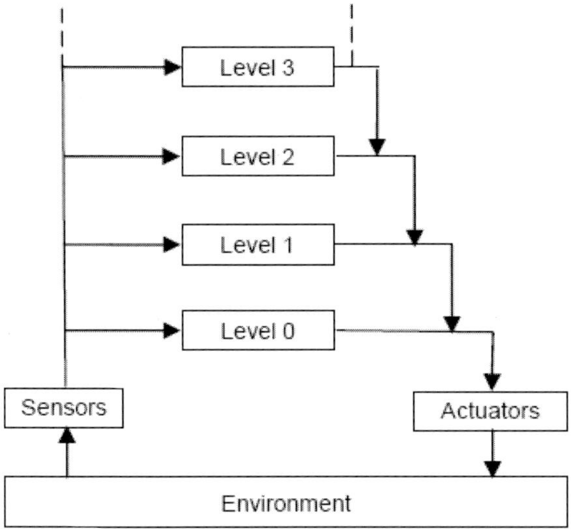

Fig. 10.7. The Subsumption cognitive architecture (Brooks, 1999).

order to achieve a higher level of competence, a new layer of control can be simply added without altering existing layers.

Functions and processes. It is notable that symbolic processing is absent in the Subsumption architecture. There is no explicit representation of knowledge; therefore, there is no matching of rules. As a result, no decision has to be made at any level in the system. Subsumption architecture is also free from central control (6, 7). The layers in the architecture are driven by data collected, with no global data or dynamic communication. This implies that the system is being reactive via implementing an activity as a consequence of events. The perception of the system would be tightly coupled to the action within each layer. The layers are also expected to react quickly in order to sense rapid changes in the environment. Communication between the layers occurs in one direction only, resulting in minimal interaction between the layers. However, as mentioned earlier, the higher layers can suppress the input and inhibit the output of the lower layers, which in turn leads to adjustment in behavior to fulfill the overall goal.

The architecture is also able to cope with noise due to imperfect sensory information or unpredictability in the environment. By designing multiple distributed layers of behavior into the system, the possibility of system collapse due to a drastic change in environment is reduced. This allows the system's performance to degrade gradually, rather than failing completely when facing non-ideal input.

The layered-based design of Subsumption architecture has also led to easier implementation. Each module of the system can be tested and debugged until flawless before proceeding to the next higher level. This type of incremental, bottom-up approach of the model ensures the workability of the system and also simplifies the debugging process. The layered design of the architecture is analogous to the biological nervous system, wherein new sections of the brain are developed for new functions, but old sections are still preserved to perform their original tasks.

CLARION Architecture

CLARION stands for Connectionist Learning with Adaptive Rule Induction ON-line. With its root in connectionist systems (i.e. neural networks), CLARION is a hybrid architecture that incorporates both implicit and explicit memories for reasoning and learning (Sun *et al.*, 2001). Procedural knowledge (implicit memory) can be gradually accumulated with repeated practice, and deals with practiced situations of minor variations. To deal with novel situations, declarative knowledge is required to guide in the exploration of new situations, thereby reducing the time for developing specific skills. It also unifies neural, reinforcement, and symbolic methods to perform on-line, bottom-up learning. Hence, CLARION is able to react in a dynamically changing environment without any preexisting knowledge installed into the architecture (Sun and Peterson, 1996, 1998).

Architecture. As shown in Fig. 10.8, the CLARION architecture consists of two levels: a top level containing prepositional rules and

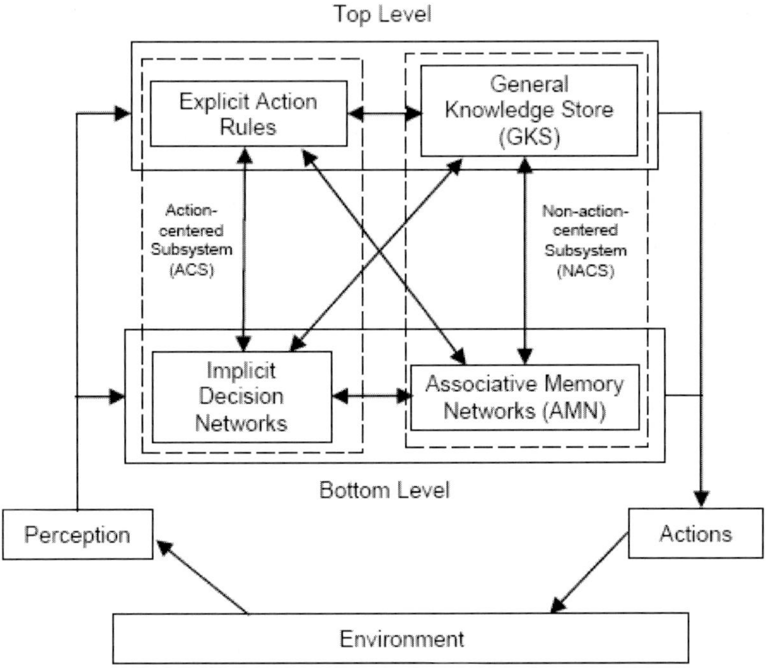

Fig. 10.8. The CLARION cognitive architecture (Sun and Zhang, 2006).

a bottom level (reactive level) containing procedural knowledge. The top level comprises of explicit symbolic mechanisms and the bottom level uses subsymbolic neural mechanisms. Procedural knowledge can be acquired through reinforcement learning in a gradual and cumulative fashion, while declarative knowledge is acquired through rule extraction by trial and error. The architecture consults both the top and bottom levels, in order to select the most appropriate action to perform at each step.

It is generally accepted that declarative knowledge is explicit, while procedural knowledge is implicit. As a result, declarative knowledge is usually accessible but procedural knowledge is not. However, there may be exceptions. Thus, the CLARION architecture includes a non-action-centered subsystem (NACS) and action-centered subsystem (ACS). NACS contains mostly declarative knowledge, while ACS contains mainly procedural knowledge. The top level of the NACS is a general knowledge store (GKS) and contains explicit representation. On the other hand, an associative memory network (AMN) forms the bottom level of NACS, which consists of implicit representation. In ACS, explicit action rules are stored in the top level. These rules are either provided by the agent's external environment or extracted from information in the bottom level. The bottom level of ACS comprises of implicit decision networks, which can be trained by reinforcement learning (Sun and Zhang, 2006).

Functions and processes. Learning in CLARION can be differentiated into implicit learning or explicit learning. Implicit learning can be considered as learning procedural skills, while explicit learning takes place by learning those rules in declarative knowledge. The learning of procedural knowledge at the bottom level occurs through a reinforcement learning paradigm, which works by making adjustments to the probability of selecting a particular action. When an action receives a positive reinforcement, the chance of selecting the action increases; otherwise, the chance of selection decreases. Learning processes employed in CLARION are through neural mechanisms, such as multi-layer neural networks and a backpropagation algorithm, which are used to compute Q-value. When the situation becomes better due to an action performed, the Q-value of the action rises, thus increasing the tendency to perform that action.

The learning of rules in the top level takes place by extracting information from the bottom level. Upon the successful execution of an action, the agent extracts a rule corresponding to action selected by the bottom level and adds this to the rule network. The agent in turn removes more specialized rules in the rule network and only keeps the general rules. The agent then tries to verify the rules learnt by trying them out in subsequent interactions with its environment. If the performed action results in an unsuccessful attempt, the rule is modified to be more specific and exclusive of the current situation. In contrast, a successful attempt enables a rule to be generalized to make it more universal in application (Sun and Peterson, 1998).

It has been shown that people often use similarities between objects in determining the outcome of inductive reasoning. As a result, CLARION includes both rule-based and similarity-based reasoning to mimic human reasoning. Reasoning takes place in CLARION via the comparison of a known chunk with another chunk. When the similarity between two chunks is sufficiently high, inference regarding the relations between them can then be made. The usage of chunks during reasoning constitutes rule-based reasoning, while the comparison of chunks makes use of similarity-based reasoning. The process of reasoning in CLARION can also occur iteratively to allow all possible conclusions to be found. In iterative reasoning, the conclusion reached at each step can be used as a starting point for the next step. The overall outcome is to select the appropriate course of action for the agent to react to its environment (Toel *et al.*, 1996).

Functional Comparison of the Six Cognitive Architectures

Comparing the six cognitive architectures across the functions is not an easy task. Each architecture focuses on different aspects of the cognition and uses a distinct set of terminology. In these comparisons, we briefly break the functions into seven functional areas, namely perception (the ability to receive information from external environment), memory, goals, problem solving, planning, reasoning/inference, and learning.

In a Soar agent, the information from the external environment is directly fed to the working memory via the perception module. The perceptual input from Soar can be from many different sources. In BDI architecture, the perceptions are available in the form of events and stored in the queue. The subsumption architecture was originally designed for robots. Thus, the system perceives the environment through obtaining data from its sensors.

Table 10.2 shows the comparison based on memory. Soar used a centralized working memory containing preferences,

Table 10.1. The Realization of Perception in the Six Cognitive Architectures

Architecture	Means of Perception
Soar	Perceptual input stored directly as part of the working memory
ACT-R	Perception stored in visual module and made available through visual buffer
ICARUS	Perceptual stored in Perceptual buffer as part of conceptual memory
BDI	Perception mapped to events stored in events queue
Subsumption	Perception is available through sensors
CLARION	Perceptual input represented as dimension/value pairs

Table 10.2. Implementation of Memory Functions in the Six Cognitive Architectures

Architecture	Representation	Working Memory	Long Term Memory
Soar	Symbolic	Contains perceptual input, states and production rules relevant to current situation	Contains procedural, declarative (semantics) and episodic memory
ACT-R	Symbolic	Contains goal, perception, relevant knowledge and motor action in the various buffers	Contains declarative knowledge in declarative module and perceptual knowledge in production system
ICARUS	Symbolic	Consists of perceptual buffer, belief memory and short term skill memory	Procedural knowledge stored in long term skill memory, declarative knowledge in long term conceptual memory
BDI	Symbolic	Belief as working memory	Plans as long term memory
Subsumption	No explicit representation	No explicit working memory	Heuristics within each layer
CLARION	Symbolic + Subsymbolic	As temporary information storage	procedural at bottom level Declarative at top level

Table 10.3. The Goal Representation of the Six Cognitive Architectures

Architecture	Goal Representation
Soar	Goals represented as states in working memory Support goal-subgoal hierarchy
ACT-R	Goals stored in the international module and made available through the goal buffer
ICARUS	Goals placed in the goal stack Support goal-subgoal hierarchy
BDI	Non-conflicting goals as desires Support goal-subgoal hierarchy
Subsumption	Multiple goals across modules in different layers Allow multiple and parallel goal processing
CLARION	Goals stored in a goal structure, such as goal stack or goal list

Table 10.4. The Comparison on Planning Capabilities in the Six Cognitive Architectures

Architecture	Machanism of Planning
Soar	Decision cycle selects appropritate actions, bringing system closer to goal
ACT-R	Planning by creating subgoals
ICARUS	Planning through instantiation of skills
BDI	Predefined plans stored in the plan library
Subsumption	Implicit planning by task decomposition
CLARION	Planning by beam search in the Q-value space

Table 10.5. Comparison of the Problem Solving Mechanism of the Six Cognitive Architectures

Architecture	Problem Solving Mechanism
Soar	Means-ends analysis Decision procedure for selecting appropriate operator to apply
ACT-R	By activation of chunks in Bayesian framework and production rule firing when chunks match with rules
ICARUS	Means-ends analysis by searching and backtracking
BDI	Means-ends analysis
Subsumption	No explicit problem solving mechanism
CLARION	Combination of Q-values calculated in the bottom level and rules in the top level to choose the course of actions

Table 10.6. Comparison on the Reasoning and Inference Capability of the Six Cognitive Architectures

Architecture	Mechanism of Reasoning/Inference
Soar	Rule matching to bring relevant knowledge to working memory
	Elaboration phase to create preference
ACT-R	Probabilistic reasoning
ICARUS	Boolean match of conceptual clauses, bottom-up and data driven
BDI	Procedural reasoning system
Subsumption	Absence of reasoning mechanism
CLARION	Integrate rule-based reasoning and similarity- based reasoning

Table 10.7. Comparison of Learning Capability

Architecture	Mechanisms of Learning
Soar	Chunking for creation of new production rules
	Reinforcement learning for updating reward of each rule
	Episodic and semantic memory to aid in decision making
ACT-R	Production compilation for reducing multiple production rules into one
CARUS	Skill learning
BDI	Q-learning, top-down induction of decision tree
	Learning from interpretations
Subsumption	Absence of learning mechanism
	Reactions to environment pre-wired into each module
CLARION	Q-learning at bottom level (procedural knowledge)
	Rule extraction at top level (declarative knowledge)

perception, states, substates, etc. Whereas ACT-R uses a distributed memory network, wherein the goals, beliefs, sensory and motor signals are situated in distinct buffers. Both ICARUS and BDI do not include procedural or declarative memory. BDI and subsumptuous architectures do not refer to any memory structure.

Goals are the objectives that the architecture agents have to achieve. All the architectures incorporate goal setting and most of them have goal-subgoal hierarchy.

Table 10.8. The Relevance of the Six Cognitive Architectures in Neurobiology

Architecture	Relevance to Neurobiology
Soar	No reference to brain anatomy
ACT-R	Used for predication of brain activation pattern; Modules and production system are mapped to various brain regions
ICARUS	No reference to brain anatomy
BDI	No reference to brain anatomy
Subsumption	No reference to brain anatomy; Layered design is analogous to biological nervous system
CLARION	Based on neural networks but no reference to brain anatomy

Conclusions

From the US DARPA BICA (Biological-Inspired Cognitive Architecture) Phase 1 proposal, there emerged a few more cognitive architectures. These are briefly described in the notes for Chapter 10. However, DARPA has silently ended the BICA project (The Star-Ledger News, 15 March 2007, "A quiet death for the bold project to map the mind" by Kelvin Coughlin).

Current cognitive architecture is mainly based on:

- Rule-based technique, i.e. the techniques are pretty much structured based on rules. Rules could come in the form of if-then statements or through matching the closest by some kind of template and then selecting the closest one as output. Although rules can be generated automatically, current auto-generation of robust rules can be improved.
- Symbolic. The knowledge representations are mainly symbolic.

The architectures have common memory systems such as long-term memory, short-term memory, an emotion component, and reasoning components. Some architecture such as SOAR and ACT-R are more detailed and include procedural, episodic, and semantic memories.

Currently, there is no single cognitive architecture or theory that can cover any significant fraction of the phenomena of human cognition and the brain as a whole. Most of the researchers in cognitive architecture design indicate that their design is still incomplete in many details and only map onto the gross anatomy and functionality of the human brain.

Chapter 11

ARE WE THERE? WHAT CAN THE COMPUTER DO TODAY AND TOMORROW?

What can the computer do now? The term computer here means any device that involves a computational process, e.g. a robot, computer system (desk-top or laptop), and other gadgets that use processors.

Data and Communication Capability

Can a computer search for data? Yes. It could probably search faster than a human in a sequential or brute-force way where the data are properly tagged, or where the search involves the use of pure key word matching.

Can a computer communicate across mediums such as a messaging? Yes, we have seen computers send text messages automatically and most of the time we are annoyed by the advertising messages.

Can a computer speak with a human-like voice? Not quite. It still has a distinct "machine" voice (Olive, 1996). For example, the ALICE has a distinct computer voice (http://www.alicebot.org/join.html).

Can a computer translate across languages? Yes, but it still not as good as a human translator in terms of accuracy. There are a number of translation engines available on the web that one can try out.

Can a computer transcribe print? Yes, for typewritten text. Software for optical character recognition (OCR) are available. OCR can translate images of typewritten or handwritten text into machine readable text. Typewritten text has reached an accuracy of 99% and is largely considered a solved problem. Research is now on handwritten text. OCR ability to recognize handwritten text is still a challenging issue. Some researchers have claimed their intelligent character recognition (ICR, another way to call OCR) is able to achieve an accuracy of 80%–90% for "clean neat handwriting".

Can a computer perform speech-to-text translation? Yes. Today, speech to text translation software claims to be as good as 99% accurate after training. The speech recognition software under noisy environments looks pretty good too. There are a number of software that can be easily checked out from the internet. (Humans may not be able to perform this task as fast as computers.)

Can a computer communicate emotions via voice/facial emotions? Voice synthesizers will produce various tones according to punctuations such as the exclamation mark (!) or question mark (?). However, this is still not quite there yet.

Can a computer understand instructions? Only simple ones. (And the instructions must have been taught beforehand.)

Physical Capability[a]

Can a robot play ping-pong? Yes. Russell L. Andersson built one robot ping-pong player (Andersson, 1988).

Can a robot run and walk like a human? Not quite yet. Sony QRIO became the first robot to run in 2003. QRIO can also dance, recover from falls, and balance on a skateboard. In 2004, a Honda robot named ASIMO (see Fig. 11.1) claimed to be running four times faster than Sony QRIO. ASIMO was running at 3 km/hour, which is closer to a leisurely jog.

---

[a] The term "robot" is used here in place of "computer".

Fig. 11.1. Honda's robot ASIMO running. Picture from http://asimo.honda.com/

More video clips of ASIMO's other physical capabilities such as climbing staircases, kicking balls, and dancing can be found at http://asimo.honda.com/asimotv/.

In 2005, researchers from the Communications and Cybernetic Research Institute of Nantes in France and the University of Michigan claimed to have developed mathematical principals for enabling human-like running and walking with the ability to recover balance. Their robot, named 'Rabbit', can walk at an average speed of 5 km/h and run at 12 km/h.

At about the same time (2005/2006), Albert Hubo from Humanoid Robot Research Center (Hubo Lab from Korea Advanced Institute of Science and Technology) also demonstrated that their robot can walk and perform some simple dance movements.

French humanoid robot, called Nao (built by French start-up Aldebaran Robotics), tried to compete with advanced domestic humanoid robots from Asia. The Linux-based, Wi-Fi equipped robot can walk, dance, understand voice commands, recognize faces, and have some 'user-teachable' capabilities.

Germany's University of Gottingen has reported progress in developing RunBot, a robot that can adapt to different terrains and adjust its posture after a few learning experiences.

The physical ability of robots to move, walk, and run over bumpy or hilly terrains is still a challenge. Humans, in general, still have a far better biomechanical system (leg, hips, etc.) and more complex neuronal control of their body system. And currently, the robotic physical ability does not look natural. This is partly due to the control mechanism and the lack of good morphology and material in the physical design.

Vision Capability

Does a computer have invariant object recognition capability, such as recognizing CAPTCHA (**C**ompletely **A**utomated **P**ublic **T**uring Test to tell **C**omputers and **H**uman **A**part)? Not yet. Humans still outperform the computer. Since humans are so good at object recognition, they are used to label the vast quantity of images in the internet through a fun game called the ESP game. The basic idea of the ESP game is to display an image and guess what label the other player will use on the image. So the usual method to guess correctly is to type as many words that describe the images and hope that one of them will match the other player's. In this way, you get all the images labeled for free by playing the game (Luis von Ahn from Computer Science Department at Carnegie Mellon University, 2006, Google technical talk 26 July 2006).

Artistic Capability

Can a machine paint a picture that looks like a painting by a human? Yes.

Can a computer create art? The robot, Aaron, created by Harold Cohen, mixes paints, chooses a subject, and paints a picture. Harold Cohen, an emeritus art professor at the University of California at San Diego said that the creative part of Aaron is the software. Another art painting computer is the robot, Action Jackson. The creator of Action Jackson, Topher McFarland, acknowledged that Action Jackson is just a mechanical plotter carrying out his instructions (instructions written in the software code downloaded to the robot).

Can a computer sing? Yes. In 1962, John L. Kelly and Carol C. Lochbaum employed a synthesis-by-rule algorithm to have a computer sing the song, "Daisy, Daisy". See also VocalSynthesis: Cantor by MacMax.

Can a computer compose music? Yes. In 1997, Steve Larson, a University of Oregon music professor arranged a musical variation of the Turing test by having an audience attempt to determine which of three pieces of music had been written by a computer and which of the three had been written two centuries ago by a human named Johann Sebastian Bach. Larson was only slightly insulted when the audience voted that his own piece was a computer composition, but he felt somewhat vindicated when the audience selected the piece written by a computer program named EMI (Experiments in Musical Intelligence) to be the authentic Bach composition (Kurzweil, 1999).

Mental Capability

Can a computer play chess? Yes, we saw IBM Deep Blue beat world champion Gary Kasparov in May 1997.

Can a computer prove mathematical theorems? Perhaps not quite. But in 1996, the theorem-proving program (EQP), developed at Argonne National Laboratory, proved the Robbins conjecture — a previously unsolved problem in Boolean algebra.

Can a machine navigate faster than humans? Yes, today, the GPS-guided navigational systems are pretty robust, fast, and accurate in recalculating alternative routes when a human makes a wrong turn. With such systems and a voice recognition system installed, driving on the highway or along the countryside becomes a breeze.

Fig. 11.2. Picture of Kismet showing a happy expression.

Does a computer have emotions? Not currently. There has been research on building machines with affective learning capabilities, and machines may one day "look like" they have emotions. For example, in MIT, robot Kismet (see Fig. 11.2) is designed to have some "human-like" expressions. Kismet can "express" emotions such as happiness, sadness, interest, disgust, surprise, anger, and calmness. For more information on Kismet, refer to the following website: http://www.ai.mit.edu/projects/humanoid-robotics-group/kismet/kismet.html.

Kismet is equipped with visual, auditory, and proprioceptive sensory input. Picture taken from http://www.ai.mit.edu/projects/humanoid-robotics-group/kismet/kismet.html.

Can a computer understand what it is saying? Not quite.

Can a computer understand what you are saying? No. Maybe it is easier for you to type into the computer for it to "understand" what you are saying.

Does a computer have mental vision, i.e. foresight? Not currently.

Can a computer understand the emotional state of another human? Not currently.

Does a computer have feelings? No, we do not fully understand how our brain generates feelings.

Can a computer teach, adapting to a student's style of learning? Not yet.

General Task Capability[b]

Can a computer discuss art, science, technology, and history? Not currently. Computer is still not able to answer questions in a proper way (given the free text knowledge), let alone striking a discussion with a human. With advances in ontology and knowledge representation research, a computer may be able to answer well structured questions which it has rich ontological representation.

Can a computer have personal experiences and discuss social relationships? Not at the current stage.

Can a computer, in general, learn to play all kinds of games such as ping-pong, badminton, golf, and soccer? Not at the current stage. Can all humans play all kinds of games?

Can a computer think like a human? (Lee, 2006) Still far from it.

Does a computer have language understanding? Not yet. Computer language understanding remains a complex task (Olive, 1996).

Can a machine demonstrate creativity? Not currently.

Can computer vision match the human capability in recognizing objects in various spatial orientations, size, luminance, etc? Not at the current stage.

Can a computer exhibit free will? Not as human free will. However, if we start to define free will, we will be confining the free will within that definition and the machine may one day achieve it within that confined definition.

The greater the freedom of a machine, the more it will need moral standards. I do not think designers will easily be able to enforce "The Three Laws"[c] for computers, since computer perception of potential harmful events to people is not perfect, and since

---

[b] General Intelligence, such as the ability to plan, reason, learn, solve problems, deal with novel situations, fuse and interpret sensory data.

[c] See notes on Chapter 14 for a definition of the three laws.

ultimately the laws are based on the machine's judgment about a situation. Furthermore, computers can be hacked, by humans or other computers. A system that truly operates in a complex and unpredictable environment will need more than laws; it will essentially need values and principles, a moral compass for guidance, and perhaps even religion (Picard, 1997).

Can a computer have a character? Still uncertain.

Can a computer be social and interact with human? Not quite at the current stage.

Can a computer deal with surprises? Not currently. Computers are good at tasks that are pre-programmed, i.e. as anticipated by the designer of the system. Human brains can react and handle surprises in real time that today's computers still cannot. The DARPA's brain-on-a-chip venture, called "Systems of Neuromorphic Adaptive Plastic Scalable Electronics" (SyNAPSE) is said to seek development of a brain-inspired electronic "chip" that mimics the function, size, and power consumption of the biological cortex. And one of the tasks is to be able to deal with surprises. (See http://blogs.spectrum.ieee.org/tech_talk/2008/11/darpas_synapse_seat_of_your_pa.html)

Could a machine be self-generative? For example, a machine builds another machine (like giving birth to a new machine). Current multiple robots for car production plants are all preprogrammed and fixed in their production process.

Could a Machine Pass the Turing Test?

How soon could we create a machine with full human cognitive intelligence?

What is the Turing Test? In 1950, Alan Turing published a paper that described his concept of the Turing Test, in which one or more human judges interviews computers and a human using terminals (so that the judges won't be prejudiced against the computers for lacking a human appearance) (Feigenbaum and Feldman, 1963). The nature of the dialogue between the human judges and the candidates (i.e. the computers and the human) is similar to an online chat using instant messaging. The computers as well as the human try to convince the human judges of their humanness. If the human judges are unable to reliably unmask the computers (as imposter humans) then the computer is considered to have demonstrated human-level intelligence.

Turing was very specifically nonspecific about many aspects of how to administer the test. He did not specify many key details, such as the duration of the interrogation and the sophistication of the human judge and human player. (Information from the Association for Uncertainty in AI (http://www.auai.org/).)

What is Alan Turing's own prediction? The common understanding is 50 years from 1950, which in this case would be the year 2000. But Jack Copeland thinks this is an unofficial prediction date of Alan Turning. Jack Copeland shared Turing's 1952 conversation between Newsman and Alan Turing, which indicated that Alan Turing said it would take at least 100 years to pass the Turing test (Copeland, 2006):

Newsman: "*I should like to be there when your match between a man and a machine takes place, and perhaps to try my hand at making up some of the questions. But that will be a long time from now, if the machine is to stand any chance with no questions barred.*"

Alan Turing: "*Oh yes, at least 100 years, I should say.*"

What do people say?

There are many diverse views on how soon the computer can pass the Turing test. Currently, on the web, there are two well-known competitions to beat or bet on the Turing test. These are:

- The Loebner Prize in Artificial Intelligence — The first Turing Test Competition. (http://www.loebner.net/Prizef/loebner-prize.html.) Hugh Loebner, a businessman, funded a US$100,000 prize for the first computer that can pass the test, and annual contests are held with smaller prizes. The 2007 contest was held in New York City. ALICE was the winner of the 2004 Loebner Prize.

- AAAI competition — Long bet between Mitchell Kapor and Raymond Kurzweil on the title: By 2029 no computer — or "machine intelligence" — will have passed the Turing Test. The vote, as of May 2007, was 53% disagree and 47% agree. Check out: http://www.longbets.org/bets (What is your vote?).

Now and then, there are also debates on this topic, such as the debate in MIT Computer Science and Artificial Intelligence Laboratory (CSAIL) on 30 November 2006 (Simon Lecture) between David Gelernter and Raymond Kurzweil under the seminar title: Creativity: the mind, machines, and mathematics.

> "*... we have to attack the problem right .....Building machine with consciousness out of the software is impossible ...the only way currently I know to achieve consciousness is through living organism ...*" — David Gelernter's reply to Raymond Kurzweil with regards to building a conscious machine.

> "*...Exponential power of growth in the information technology, which will affect hardware and software ... we will have model and simulation of the various cognitive regions in the brain within 20 years ... and that will give us template to build intelligent machine to pass the Turing test.*" — One of many remarks by Raymond Kurzweil on why he thinks the computer will beat the Turing test within 20 years.

> "*The chance to pass Turing test in the next 50 years is good.*" — Copeland's concluding remark in Simon lecture on 30 Nov 2006, MIT/CSAIL.

> "*Within 25 years, we'll reverse-engineer the brain and go on to develop superintelligence. Extrapolating the exponential growth of*

> *computational capacity (a factor of at least 1000 per decade), we'll expand inward to the fine forces, such as strings and quarks, and outward. Assuming we could overcome the speed of light limitation, within 300 years we would saturate the whole universe with our intelligence.*" — Raymond Kurzweil (The Intelligent Universe).

Raymond Kurzweil indicated that with an exponentially increasing computational ability provided by the Law of Accelerating Returns (a variation on Moore's law), we will be able to continue to rapidly improve the resolution and speed of non-invasive brain scanning technologies and enable humans to build a new "brain".

Note also that the Law of Accelerating Returns can only be achievable when quantum computing can be achieved. What is today's state-of-the–art in quantum computing? Are we anywhere near a working model?

On April 13, 2005, Gordon Moore himself stated that the law may not hold for too long since transistors may reach the limits of miniaturization at atomic levels.

> "*In terms of size (of transistor) you can see that we're approaching the size of atoms, which is a fundamental barrier, but it'll be two or three generations before we get that far — but that's as far out as we've ever been able to see. We have another 10 to 20 years before we reach a fundamental limit. By then they'll be able to make bigger chips and have transistor budgets in the billions.*" — Gordon Moore (Dubash, 2005).

The task ahead to increase computational speed is challenging and this may have a potential impact on how much we could reverse engineer the brain, unless alternative methods are available.

Roger Penrose in his book, "The Emperor's New Mind", argues that there are facets of human thinking and human imagination that can never be emulated by a machine. Penrose indicated that laws, even more wondrously complex than those of quantum mechanics, are essential for the operation of a mind (Penrose, 1991).

> "*I believe that the creation of a superhumanly intelligent AI system is possible within 10 years, and maybe even within a lesser period of time (three to five years). Predicting the exact number of years is not possible at this stage. But the point is, I believe that I have arrived at a detailed software design that is capable of giving rise to intelligence at the human level and beyond.*" — Ben Goertzel in Artificial General Intelligence: Now Is the Time, published on April 9, 2007.

> "*Why have we made limited progress in AI? Because we haven't developed sophisticated models of thinking, we need better programming languages and architectures, and we haven't focused on common sense problems that every normal child can solve.*" — Marvin Minsky (The emotion universe) (Minsky in a 2007 class commented that many people have proposed general AI but they all get "flattened out". Perhaps building general artificial intelligence (GAI) is more complex than we have imagined.

> "*The reason that no computer program can ever be a mind is simply that a computer program is only syntactical, and minds are more than syntactical. Minds are semantical, in the sense that they have more than a formal structure, they have a content.*" — John Searle, Minds, Brains and Science (Searle, 1984)

Hence, our current approach of modeling the biological plausibility is just a small fraction of the true biological network. Yet, in this small fraction we can see some impressive results although it is still far from a general intelligent system. Nevertheless, the quest continues and more funds are allocated to studying the brain with the aim to build general intelligent systems. These include US DARPA's initiative on a number of brain-inspired computing projects such as project SyNAPSE. The European Commission's 7th Framework Programme listed "Cognitive systems, Interaction, Robotics" as one of its key challenges. A number of research institutions and laboratories are doing research on computational cognitive science areas with the goal of building general intelligent systems.

Chapter 12

BRAIN — A FOREST NOT TOTALLY EXPLORED: WHAT ARE SOME OF THE ISSUES?

We now have a lot more data about the brain then a decade ago, yet we still cannot explain many things that happen in the brain.

Today, the "forest" in our brain is not a totally unexplored subject. The forest has been "probed", "measured", and observed from physiological and behavioral data. However, there are still many issues that can be researched on and many questions that need to be resolved. Many scientists have provided many interesting views, opinions, and suggestions of how the different parts in the brain work. There are also many different claims of why and how the brain, as a whole, works. These claims are mainly based on data collected from brain measuring instruments.

This chapter presents some of these claims and the issues to address:

- The different theories and views of how the brain and mind work
- Some open issues
- Some limitations of the instruments.

Many Theories and Views of How the Brain and Mind Work

In the mid-1960s, the brain was likened to a telephone switching network; in the mid-1980s, it was compared to a digital computer. By the end of 1990s, it is said to be similar to a massively parallel digital computer. More recently, the brain is compared to a quantum computer as well as to the World Wide Web.

As discussed in the earlier chapters, at the fundamental level, the brain consists of many nerve cells (neurons). The nerve cells interconnect and associate with one another throughout the nervous system. These form the fundamental parts of the brain that give rise to memory and learning. There are many mysteries and wonders that go on in these neurons and their interactions. Many scientists have different ideas, concepts, and theories of explaining these massive neurons, their highly distributed and parallel interaction among one another that lead to a working mind. From their professional points of view, whether from sciences (such as chemists, biologists, or physicists), engineering, computer science, neuroscience, psychology or philosophy study, they all share valuable insights into this great mystery and wonder of how the brain and mind work. Here are some of their insights.

Brain as an associative network. The neurons' massive interconnection has led many to propose that the brain is organized in an associative network and governed by some rules such as reinforcement or associative rules.

Brain as having many distributed functions. The late Professor Karl Lashley (1890–1958) described the brain as having many functions. Each of these functions is not localized but distributed around the brain. Every brain region partakes (to some extent) in all brain processes. For example, the function of memory in the brain is not localized but distributed throughout the cortex.

Brain as a dynamical system. Some scientists hypothesize that the electrical activities of the brain can be described by the theory of dynamical system (van Gelder 1998; Laurent *et al.*, 2001). Progressing from this thinking, the following have been arrived at:

- Chaotic dynamical system. The brain as a chaotic dynamical system (Skarda and Freeman, 1987; Tsuda, 2001; Kay, 2003).
- The brain as a network of oscillators. Neurons are oscillators and they modulate other cells or networks of cells. Various messenger chemicals, such as peptides, modulate the effects of other transmitters. And a significant population of neurons rarely or never "fire" (Bullock, 1981; Roberts and Bush, 1981). Instead, they propagate slow-wave graded ionic potentials (Bickhard and Terveen, 1995).

Professor Stephen Grossberg's shunting networks can be said to be a class of dynamical system, which uses first order differential equations in the model.

> "*Brains are open thermodynamic systems, continually dissipating metabolic energy as heat in constructing spatiotemporal patterns of neural activity. Here, patterns of cortical oscillations are described as Prigogine's 'dissipative structures' formed at a conditionally stable operating point far from equilibrium.*" — Walter Freeman (Freeman, 2007) (Freeman, 1997).

Professor Mark Bickhard from Lehigh Univeristy suggested that the brain is composed of many oscillators that generate vast concurrent micro-modes (Mark called these micro-modes a microgenesis process) of processing across the brain. These oscillators are neurons that oscillate and resonate with modulation where time places a part. And it could anticipate whether a stimulus it received is true or false in the environment. The microgenesis process anticipates that the appropriate condition — whatever it is — obtained is the current environment.

> "*An architecture supporting oscillations and modulations makes very strong connection with a number of aspects of brain functioning that both connectionist models and standard models ignore.*" — (Bickhard and Terveen 1995).

Brain as composed of multiple intelligences. Professor Howard Gardner, from his study of the brain in an effort to ascertain the optimal taxonomy of human capacities, coined the term multiple intelligences. He indicated that human beings are best described as having a set of relatively autonomous intelligences such as linguistic, logical, spatial, bodily kinesthetic, musical, interpersonal, and intrapersonal intelligences (Gardner, 2003).

Brain as a map machine. Experiments have shown that in the mammalian visual cortex, neurons are organized according to their functional properties into multiple maps such as retinotopic, ocular dominance, orientation preference, direction motion, and others. "Map", here, refers to a pattern or representation of a sensory surface (such as the retina or skin receptors, which receive the stimulus) onto the cortex's neural surface. Some believe that these cortical maps reflect neuronal connectivity in the brain. In the 1970s, (Schwartz, 1977) indicated that the retinotopic mapping of the visual field to the surface of the striate cortex is characterized as a logarithmic mapping. This was subsequently observed in human subjects using a positron emission tomography (PET) scanner following the administration of 2-deoxyglucose (2DG) to human subjects, and a logarithmically structured "ring-ray" stimulus was observed (Schwartz *et al.*, 1984). Hence, from the visual cortex study, it was generalized that the brain is a map machine (Schwartz *et al.*, 1988) (Chklovskii and Koulakov 2004) (Polimeni *et al.*, 2006). The study to model the cortical surface (via mapping) was useful because it is believed that the local functional organization of the cortex is largely 2-D.

Brain as an information processing system. This has been the one of the key themes of this book, which looks at the brain as an information processing system, and to obtain the inspiration for building a computationally intelligent information processing system. The brain has a network of neurons (neural networks) that forms a massively parallel information processing system. Most computational scientists take this perspective.

Our brain is complex and can be described and viewed in many different interesting perspectives. There are also many interesting different hypotheses on how the mechanism of the neurons and synapses work that give rise to a working mind. The matter on the brain is clearly unsettled and fuelled by many challenging research issues.

Open Issues

What are the open issues? There are many open issues with regards to the study of the brain. No one can fully claim that we have now understood how the brain works and how it gives rise to a mind. The exact cortical connection is probably far more complicated then it is thought to be. Many open research issues remain with regard to how the neurons and group of neurons are interrelated and connected. From the molecular level to the functional region level there remain many issues for research. As the problem is addressed by researchers from diverse disciplinary fields and different cultural backgrounds, there are also communication gaps such as the understanding of terminology and interpretation of findings that may potentially hinder some of research progress.

A number of open research issues were discussed in the previous chapters, a few more will be discussed here, and more are provided in the Notes for Chapter 12.

What are the biochemical and molecular mechanisms that activate the different parts of the brain regions and create different functional processes? Based on laboratory tests, we know a number of roles and functions in different parts of the brain. For example, we know the role of the hippocampus in spatial memory, the amygdala in emotion, the associative cortex in decision-making, the thalamus for regulation of body temperature, etc. However, we know little of the biochemical and molecular mechanisms that activate this process and how they interact to give rise to a functional system.

"*Brain circuitry more precise than suspected*", a phrase borrowed from the SALK Institute for Biological Studies press release in 2005. One key principle in brain organization is that nerve cells (neurons) with similar functions are grouped together and organized into functional slices known as functional columns (Hubel and Wiesel, 1962; Mountcastle 1997). Each column generally carries about a thousand neurons. However, why is there a need for so many neurons to be located in the same area of the brain to carry out the same function? Neuroscientists are puzzled by this question. The latest finding by Professor Edward Callaway from the SALK Institute indicated that the thousand neurons in these columns are not the same. "*There are fine-scale connections with the columns so that different neurons right next to each other could be involved in very different functions because they are connected differently and don't even talk to each other directly. Right now, people are placing electrodes in the brain and recording from 100 neurons simultaneously and it turns out this is relatively uninformative. That's probably because most of them are not connected or even communicating with each other.*" — Edward Callaway. The brain's circuitry is organized on a much finer scale than previously suspected, creating new challenges for future studies (SALK Institute press release, 2005).

Tracing the "wires" of the brain is non-trivial. The thinnest axons are about 100 nm in diameter and it is necessary to image the structure of the brain at nanoscale resolution to trace the wires of the brain at both the dendritic and axon level. Special instrument and techniques are needed to perform this research (see notes for Chapter 12 for some details).

How are the various regions interconnected and related? Studies by (Bulkin and Groh, 2006) have suggested possible multisensory processing in the brain regions. The interrelation of the various regions of the brains is not well understood to model it effectively.

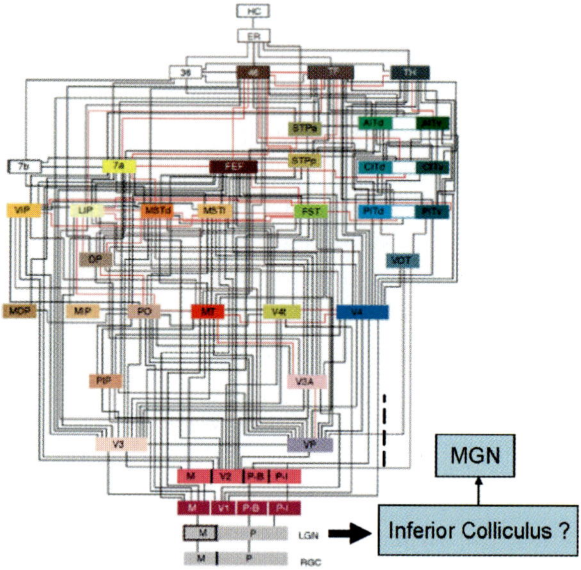

Fig. 12.1. A visual circuitry with a link to the auditory pathway.

One example is the well-known visual circuitry diagram (see Fig. 12.1), where some studies have suggested a possible link to the auditory pathway via the inferior colliculus to the parietal cortex.

> "*How do areas of the brain communicate? For example, a simple activity like responding to a question involves areas all over the brain that hear the sound, analyze it, extract the relevant information, formulate a response, and then coordinate your lips and mouth to speak. We have no idea how information moves between these areas.*" — Dr Robert Knight (Sanders, 2006).

"*It's an open question whether the cortex is really as modular as many fMRI studies make it out to be.*" — David Cox (Massachusetts Institute of Technology (MIT)) and Robert Savoy (Massachusetts General Hospital (MGH)). They recently described a new fMRI application of a statistical methodology aimed at identifying, not ignoring, the interactions among voxels from different parts of the brain (Cox and Savoy, 2003).

Problem with the precise and accurate localization of the brain region. The brain has a complicated surface of many folds (gyri) and fissures (sulci). There are some variations in shape and size of the gyri and sulci of each person's brain. This makes it difficult to compare brains across subjects to determine significant regions of function and activation, the purpose being to study the functional architectures and neural maps embedded in the brain.

To draw good inference from two independent studies, one needs to be able to accurately pin down the region of the brain studied. What constitutes the same brain region from individual to individual? Currently, there are still issues with regards to the accurate localization of which brain region performs what function for an average brain. There are four possible different methods of detecting sameness-in-regions of the brain. These are:

1. Normalized to 3-D Coordinates, i.e. transforming the subject's brain to a standardized coordinate.
2. Sulcal and Gyral landmarks, i.e. considering the subject's brain curvature and shape, localize based on prominent sulcal and Gyral landmarks.
3. Cytoarchitecture (cell types, brain's layer, size, etc), i.e. characterize brain region via cytoarchitecture.
4. Create flat maps of the human brain. The surface of the brain, i.e. the grey matter is where the functional activation of the brain occurs, and one way of visualizing the activity of the brain is to create a surface-based map (Hurdal *et al.*, 2001).

However, the problem is that the data from these four methods do not agree with one another. There are still some variants from method to method. Some have suggested methods, such as "Functional Regions of Interest", involving anatomy and function, to reconcile the difference.

Limitation of the measuring instrument. Today's measuring instruments have provided far more insight to scientists in understanding the brain than a century ago. Some of the more recent enhancements to the instruments, such as the fMRI, may have improved the research a great deal. However, this is not to say that the instruments are perfect. It is important to know the limitations so that the data collected can be taken into the perspective of the instrument's limitation. Data collected from the instrument needs to be properly processed for correct interpretation. As this is an important issue, the next section is devoted to a full discussion.

Measuring Instruments in Neuroscience

Today's measuring devices may not be as perfect as we would desire to explore the brain, particularly for the study of the brain's activity, its connectivity, or the dynamic neural correlation with the cognitive processes in the human brain. From these measuring devices, huge data have been collected. There are also issues with regards to the ability in using the data in an accurate, precise, and meaningful way for scientific discussions in exploring the brain's connectivity.

The purpose of this section is to explore the limitations and issues with regards to the measuring instruments and the data collected from these instruments.

Instruments Measuring the Brain's Electric Signal

EEG (Electroencephalography). This method measures the brain's electrical activity (potentials) by taking the record from electrodes placed on the scalp (see Fig. 12.2). The resulting traces, which are a continuous measure of the brain activity, are known as EEG and it represents an electrical signal from a large number of neurons. These are sometimes called brainwaves.

ERP (*Event Related Potentials*). ERP is a specific case of EEG recording. ERP recordings link to the occurrence of an

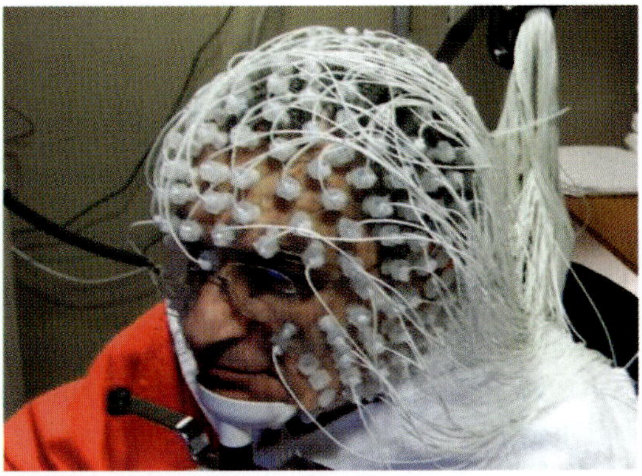

Fig. 12.2. A man's head with the EEG electrode. (Courtesy of Sekuler's lab (Brandeis University), with permission from Yigal Agam (March 22, 2007).)

event, i.e. a presentation of a stimulus would be such an event, and the brain's response to this event (stimulus) are recorded. These records in some sense represent the signal generated by thousands or more neurons in the brain responding to the event. By measuring the brain's response to an event we can learn how different types of information are processed.

The ERP's electric traces shows interest for researchers because different components of the response indicate different aspects of cognitive processing. For example, presenting different sentences to a subject will give different electrical signal responses, which will determine the time taken to register information about a word's meaning. And one can also figure out where this activity occurs in the brain. However, localization using EEG/ERP is not very accurate (poor spatial resolution).

The advantages are that the measuring of EEG/ERP is inexpensive and non-invasive. EEG and ERP have a high temporal resolution. The limitations of both EEG and ERP are as follows:

- Poor spatial resolution. The electrodes placed on the scalp could pick up useful electric signals from an area (from a large group of neurons) but not from individual neuron action potentials. And even for that area, it is difficult to pinpoint the exert location in a 3-D brain.

- Neuron activities. It is also difficult to discern from EEG/ERP signals whether the neuron activity is releasing inhibitory, excitatory, or modulatory neurotransmitters. This is because the signals emitted could be affected by cerebral tissues, fluid, and other unknown factors.
- Background noise. As in all measurements, the careful removing of background noise is needed (in this case it is the 60 Hz electrical background noise) and also the blinking of eye or movements of body muscle near the electrode may affect the electrode measurement, since the useful electric signal generated from the brain is in the order of about 50 microvolt.

  - Brain noise. Even if all the instrument-measuring noises are removed, there is also brain noise, such as random crackling and sputtering of neurons that usually go unnoticed. Furthermore, other parts of the brain may be functioning and generating noises that may not be related to the spot of interest, which the instrument is measuring.

Some techniques to overcome the limitation include:

- Using a large number of electrodes to triangulate the source of electrical activity — but it should be noted that the accuracy or required resolution is still not as good although triangulation helps.
- Removing the noise from eye blinking and power line noise. There are a number of papers that claim to automate the removal of eye blinking artifacts of EEG, e.g. using the Independent Component Analysis (ICA) (Li *et al.*, 2006) and the Constrained Blind Source Separation algorithm (CBSS) (Shoker *et al.*, 2004).

However, all the techniques for overcoming the shortfall are far from perfect at the current stage, and hence, one's claim of the experimental results using EEG/ERP data should have careful consideration of the limitation.

Instruments Using the Magnetic Field

MRI (Magnetic resonance imaging). MRI was discovered in the 1930s by Felix Bloch and Edward Purcell. In the 1970s, MRI started to be used for medical diagnosis. Then, it became one of the methods that provide a better picture of what is inside the skull or the brain. It can be used to determine brain size and employs a strong magnetic field. Currently, there are no known risks or side effects. However, any metal implant or metallic object nearby would greatly affect the magnetic field reading.

In the MRI-related family, we have fMRI and diffusion MRI.

fMRI (functional Magnetic Resonance Imaging). The most commonly used approach for fMRI techniques in brain imaging was introduced around 1992 (Kwong *et al.*, 1992). Noninvasive techniques (advantage point) are used. fMRI has dominated the field of human brain mapping since the 1990s, and till today, it is regarded as one of the most valuable measuring instruments for brain research. Many papers were published based on fMRI studies. fMRI takes a picture based on the average amount of oxygen in the blood. It is known that changes in blood flow and the amount of blood oxygenation is related to the neural activity in the brain. Studies of this blood flow are known as hemodynamic.

The fMRI picture is based on several measurements and the mean results are shown. In other words, the fMRI generates maps that are a statistical picture of brain activity, with the colors changing as neuronal activity increases. The advantages are that fMRI is non-invasive and has a high spatial resolution (3 to 6 mm). Limitations of fMRI include:

- Poor temporal resolution (in order of seconds). fMRI is not good for recording neuronal events in real time.
- The technique is limited by interpretational challenges. In (Pray 2003), the article indicated statistical skills are needed to handle

the enormous amount of data even in a single scan, to prevent erroneous conclusions and understand the truth. However, other scientists also find that the challenge is in the many steps that need to be done, such as artifact removal, preprocessing, region of interest (ROI) definition, sometimes volume segmentation, inflation, and flattening, as well as statistical tests and proper correction for multiple comparisons.

- Spatial resolution is not sufficient to resolve the activity of individual neurons (normally, an average of a group of neuron activities = approximately 20 mm^2 or more (Smith *et al.*, 2001) claims that the fMRI can take up to 16 mm^2). However, this spatial resolution is still far better than the EEG/ERP method. Nevertheless, there is still a question of what we are measuring given the spatial resolution.

    - As Ugurbil *et al.* indicated that fMRI maps are based on secondary metabolic and hemodynamic events (hemodynamics refers to the study of the properties and flow of blood) that follow neuronal activity and not on the electrical activity itself, it remains mostly unclear what the spatial specificity of fMRI is, i.e. how accurate are the maps generated by fMRI as compared with the actual sites of neuronal activity (Ugurbil *et al.*, 2002).

- Lastly, the nature of the link between fMRI signals and processes that define neuronal signaling, such as action potentials or neurotransmitter release, is not fully understood. Professor Kim and colleagues' study on BOLD fMRI (BOLD–blood oxygenation level-dependent) signals to correlate the neuronal activity have reached the conclusion that the BOLD signals are a robust predictor for neuronal activity only at supra-millimeter spatial scales, i.e. 2×2 mm^2 (Kim *et al.*, 2004). Another report by (Logothetis and Wandell, 2004) indicated that the BOLD contrast mechanism reflects changes in cerebral blood volume, cerebral blood flow, and oxygen consumption. The interaction between neural activity and these

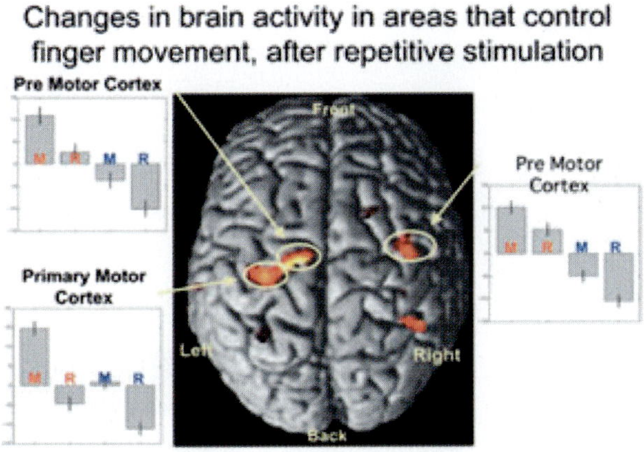

Fig. 12.3. A fMRI image showing the increase in brain activity (highlighted in color) at the site of stimulation (primary motor cortex) and in other areas of the brain that control movement (such as the premotor cortex) (Pray, 2003).

variables involves a number of factors, including the cell types and circuitry driven during activation, and the processes that couple energy demand to its supply to the brain.

A conventional fMRI scanner (see Fig. 12.4) contains giant cylin drical supercooled magnets, which produce a field some 50 thousand times stronger than the Earth's magnetic field and can create 3-D maps of the body tissue. Subjects lie on a sliding table, which is pushed into the scanner. Subjects are usually restrained with soft pads to prevent movements that might result in unusable data. Because of the strength of the magnetic field involved, subjects are required to remove all metallic objects from their body before being placed in the scanner.

Diffusion MRI (including diffusion tensor imaging (DTI)). Diffusion tensor imaging (DTI is also known as diffusion tensor MRI) was introduced by Peter Basser and his colleagues in 1994. Currently, there has been an increase in reports of researchers using DTI, or more generally, the diffusion MRI to examine different brain regions such as white matter changes in normal aging and in AD. This technique enables the measurement of the restricted diffusion

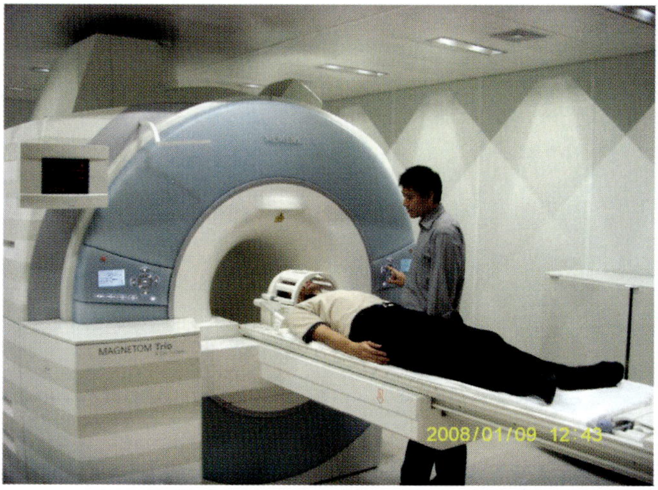

Fig. 12.4. A fMRI scanner. Courtesy of Dr Michael Chee and the staff of the Cognitive Neuroscience Lab, Duke-NUS Graduate Medical School.

of water in tissue (i.e. to observe the Brownian motion of water molecules in brain tissues). The axons in parallel bundles and their myelin shield facilitate the diffusion of the water molecules along their main direction. This preferential oriented diffusion is called anisotropic diffusion. If we apply diffusion gradients (i.e. magnetic field variation in the MRI magnet) in at least six directions, it is possible to calculate, for each voxel, a tensor (where the word "tensor" mathematically means a generalized linear quantity or geometrical entity that can be expressed as a multi-dimensional array relative to a choice of basis of the particular space on which it is defined, i.e. an organized collection of numbers) of a symmetric positive definite three-by-three matrix, which describes the 3-D shape of diffusion. The fiber direction is indicated by the tensor's main eigenvector.

The diffusion MRI limitations include:

- Low spatial resolution of the image relative to the tract curvature (Lori *et al.*, 2002).
- Ambiguity between crossing and bending fibers (Basser *et al.*, 2000) (Tuch, 2004). Also, the human brain is known to have

many complex crossing patterns, and there is limitation in estimating voxel structure (including intra-voxel structure). Voxel, here, refers to the volumetric representation of brain tissues.

- Distortions in the images (Jezzard and Clare, 1999).

Some techniques to overcome this include:

- Model-based methods (some prior knowledge of the local fiber configuration is needed).
    - Multi-fiber Gaussian tensors, which model the signal as a finite number of Gaussian fibers (Tuch, 2002).
    - Spherical deconvolution techniques, which use Gaussian kernels to estimate the diffusion signal (Ozarslan *et al.*, 2004) (Tournier *et al.*, 2004). Orientation Diffusion Function (ODF).
- Model-independent
    - Q-ball imaging (Tuch *et al.*, 2005). This method seeks to reconstruct the ODF (Orientation Diffusion Function, sometimes called the Orientation Distributed Function) from the high angular resolution diffusion imaging data. This method is said to be able to resolve subvoxel structure.

MEG (Magnetoencephalography). MEG is a related method to the EEG. But instead of recording electrical potentials, it uses magnetic potentials at the scalp to index brain activity. For example, to locate a dipole, a magnetic field can be used, due to the dipole's extreme high points of intensity of the magnetic field. By using a superconducting quantum interference device, MEG can record these magnetic fields.

The advantage of MEG includes: The magnetic field is not influenced by variations such as the different thickness of brain tissues, cerebral spinal fluid, the skull, and the scalp. The strength of the magnetic field tells us how deep within the brain the source is located. And similar to EEG, it has a good temporal resolution of up to the millisecond. However, the limitations are:

- The magnetic field in the brain is 100 millionth the size of the Earths' magnetic field. Hence, highly shielded rooms are needed.
- MEG cannot detect activity of cells with certain orientations within the brain. For example, magnetic fields created by cells with long axes radial to the surface will be invisible.
- It has poor spatial resolution.

TMS (Transcranial Magnetic Stimulation). The TMS is the reverse of MEG. It has a series of coils placed on a subject's scalp. The focused magnetic field pulses, then stimulates the underlying neurons. This causes the potential of neurons to depolarize, and the neurons fire in a random pattern rather than in a coherent order. There are two types of TMS:

- Single-pulse TMS, in which stimulation is at a particular time during the performance of a task.
- Repetitive TMS (or rTMS) is TMS with pulse frequencies of $\geq$ 1Hz. rTMS gives multiple pulses. It can facilitate brain activity or disrupt activity.

One of the major advantages of TMS is that it can be used to confirm findings from the lesion method. However, the TMS coils can only stimulate regions on the surface. It is not possible to stimulate anything deeper without stimulating what is above.

Instruments Using Radioactive Material

PET (Positron emission tomography). Positron emission from radioactive nuclei was discovered in 1933 by Thibaud and Joliot. More detailed history refers to (Valk *et al.*, 2003). It is an invasive technique.

A short-lived radioactive tracer isotope is injected into the participant (usually in the blood). The radioactive tracer enters the bloodstream and goes to the brain. These molecules are unstable but become stable by producing a positive charged electron (positron). Then two electrons will terminate each other, producing energy. The two electrons form a photon of light. Any area that is metabolically active produces more photons of light, and any area that has less produces fewer photons. The tracing of where the energy was created can enable us to localize what region at which this happens.

With PET, the blood flow in the brain is mapped, i.e. it allows us to see how the brain uses molecules and provides an absolute measure of regional cerebral blood flow. The disadvantages are:

- It is invasive (involving the injection of a radioactive substance), thus it is not possible to make many scans on a subject (participant). Thus, it is not particularly recommended for use in the human brain.
- The time period taken to get an image due to the isotope's half-life is an important factor, because once time has gone by, there is valuable brain activity we did not have an image of.
- A machine called the cyclotron, which is very expensive, is required. The cyclotron is an accelerator of subatomic particles. It produces a large quantity of protons (heavy particles with an electrical positive charge).
- PET does not provide enough imaging of brain activity from one person.

SPECT (Single photon emission computed tomography) is worse off in resolution and takes longer to get an image compare to PET. SPECT is an invasive method.

Instruments Using Optics

Optical Imaging. Optical Imaging measures the changes in cortical light reflectance. It allows us to obtain information about the neural activity as well as its time source. This method uses infrared light and the detectors are composed of optic fiber bundles, which sense how the path of light is altered either through absorption or scattering. This can be used to measure the absorption of light, which is the slow signal, and deals with the concentration of chemicals in the brain. It starts after neuronal activity and stops seconds after, which is thought to reflect increased blood flow to areas engaged by tasked demands. This method is limited by its invasiveness (it can be done during surgery) and shallow penetration into the brain. (It cannot penetrate more than ~1 mm deep into the exposed cortical surface.)

NIRS (Near-infrared spectroscopy). NIRS is a noninvasive optical technique that measures changes in the hemoglobin oxygenation state in the human brain (Jobsis, 1977). The accuracy and reliability of NIRS are still controversial. This is because of the incomplete knowledge of which region in the brain is sampled by the NIR light (Hoshi, 2003).

Combination of Instruments

By using a combination of the above instruments, some of the individual instrument's disadvantages are overcome to some extent. For example:

fMRI with EEG and MEG. We know that fMRI has poor temporal resolution, and hence, it may be complemented with the use of EEG or MEG for the same experiment set.

PET with MRI or CT. PET is often used in conjunction with an MRI or CT scan through "fusion" to give a full 3-D view of an organ.

TMS with fMRI and MRI. Combining TMS with fMRI and MRI to form a 3-D brain mapping.

One of the best future hopes is still in finding the best strategies to combine these instruments, and finding better combination techniques to leverage on their respective strengths and eliminate the weakest. Some of the challenges the combination technique would need to overcome include:

- Registration error. For example, the different instrument measurements coordinate differently.
- Temporal and spatial error. For example, MEG and fMRI cannot be measured together at the same time. Some experiments for MEG and fMRI will be conducted at different times. Due to this, the same experiment may have temporal and spatial error.

Lesion Method

Brain lesions. A brain lesion is an area of the brain that is damaged in both structure and function. The method observes the change in behavior or cognitive functions, caused by brain lesions that lead to an inability to perform a particular mental function. Most lesion studies are performed with animals. There are cases for lesions in human brains (due to diseases or injury). The advantage of using brain lesions is the claim that the method gives excellent localization. However, there are also some variants one has to take note of, such as:

- Patients vary, as does the damage to their brain, and their symptoms. And even though a researcher can compose a group that seems to have damage to the same area of the brain, it isn't always the same exact area, and the size and extent varies with each person.
- Sometimes, it is not clear if the damage caused a certain area to not function or whether the damage caused a disruption of signals being sent to the area that controls the function.

The disadvantages are that the method is very invasive, you rarely see pure cases (especially in diseases), and (when using animals) whether they are valid models for humans, there are differences between animal and human brains. Other difficulties in using the lesion method may include:

- A lack of ability to observe the function by the damaged portion of the brain also can cause problems. Researchers can only see how the brain functions after the damage has occurred, and how the rest of the brain responds to it.
- Also, the role a specific part of the brain controls is not always clear. Sometimes an area of the brain can be "masked" in a way such that it is not readily connected to a certain function.
- The final, possible limitation, of this method is that it can lead to underestimation of a certain brain function. A patient may compensate their behavior by trying to perform a task differently than they would have prior to their brain damage.

(Adapted from psychology wiki farm.)

Conclusions

Today, we have collected lot of data and understood the brain better than our ancestors. However, we are also left with many unanswered questions. The brain measuring instruments have helped provide many insights to our understanding of the brain. However, researchers need to know the limitations of each instrument, taking these into consideration when they process and interpret the data, and without overemphasizing the research results beyond these limitations.

The computational modeling of the brain is shaped by the collection and interpretation of these experimental techniques using such instruments; hence, when one models the brain, it is good to keep in mind the data's assumptions and not overstate the accuracy in understanding the brain. We do not want to reach a stage where an experimental technique or modeling of the brain is so widespread, and its limitations so poorly understood, that modeling concepts based solely on these methods take root, making the incorrect ones difficult to weed out.

> "*After half a century of cognitive revolution we remain far from agreement about what cognition is and what cognition does. It was once thought that these questions could wait until the data were in. Today there is a mountain of data, but no way of making sense of it.*" — Pamela Lyon, Journal of Cognitive Process (2006) 7:11–29. "The biogenic approach to cognition".

It would be no surprise that now and then we hear or read of news of a new brain theory or new discovery that gives new insight to the brain and mind. The brain forest is still out there, and there are rooms for many possible explorations and explanations with regards to the brain. The brain remains one of the great wonders.

Chapter 13

UNDERSTANDING THE BRAIN TO BUILD INTELLIGENT SYSTEMS

We have taken inspiration from birds to build airplanes that fly. And we do not necessarily need to design airplanes that have wings that flap like birds. Likewise, we can take inspiration from the brain to design intelligent systems. And we do not necessarily have to design the neurons at the cellular or atomic level. The airplanes that fly come in all sizes and can carry far more load than any bird. Likewise, in our approach to build intelligent systems that have human-like intelligence (or better than human intelligence in certain aspects), we need not use designs that copy the structure of the brain to the minute details or have brain-like size. Studies on the brain can help us look for key design principles and take inspiration from biological cellular structures to behavior at the organismal level. The intelligent system can be built based on these key principles.

This chapter will discuss some of these key design principles, the ways in which humans and computers can augment one another, and how the computer can leverage on its computational power to overcome some of the perceived limitations of the human brain.

What are the Design Principles of the Brain?

Some key principles we can learn from the brain to inspire our building of intelligent systems are:

1. At the atomic, molecular, and cellular level:
   a. Neuron behavior. Modeling the details of neurons at the molecular or even cellular level would be too complex to achieve, and even if it were done, one would have the difficulty of determining the accuracy and precision of the model. One possible approach is to design models that produce the same effect or have some similar neural behavior or activity. For example, one may consider models like shunting networks (Grossberg, 1988), the simple spiking model (Izhikevich, 2003), and various integrate-and-fire neuron models.
   b. Excitatory and inhibitory effects.
      i. These effects influence how information is transferred, kept, and learnt between neurons.
      ii. There are more excitatory neurons than inhibitory neurons. And intelligent systems may take this design principle of more excitatory than inhibitory effects.
   c. Synaptic strength and connection. Understanding the long-term potentiation (LTP) and long-term depression (LTD) may help in the understanding and designing of information representation, storage, and association with other links in an artificial "cortex" architecture.
   d. Parallelism. Multiple activities occur simultaneously in the neurons due to many parallel interconnections among neurons. Biologically, every neuron is connected to many other neurons and each acts like a "processor". This parallelism allows the brain to "compute rapidly".[a] The design principle

[a] Biologically, the fastest neural events occur on the order of milliseconds.

of parallel connection can be adapted, i.e. many processors connected to one another in a parallel structure similar to neuron connections may help in achieving this ultra-fast computation.[b]

e. Distributed. The ways information is transmitted, retrieved, learnt, and stored are mostly in a distributed neural system. This distributed neural system can be adopted for information processing. For example:

   i. Redundancy. If one or two neurons in the group or column[c] die it does not affect the whole column's function. Information representations are also more robust in the case of accidents or the encroachment of new concepts.
   ii. A distributed nature may make it easier to access a representation from different stages of processing or easier to "recall" by allowing for multiple indices into the representation.

f. Principles of pre-wired and evolving wiring.

   i. The key components and the basic structure of the brain are mainly pre-wired or hardwired as specified by the gene.
   ii. The basic structures are specialized through exposure to external stimulus.

g. Feedback and feedforward connections between neurons. These ideas already exist in many artificial neural networks and smart systems, where

   i. Feedback provides top-down expectation.
   ii. Feedforward for bottom-up influence.

---

[b] Parallelism does not increase the speed of the computation itself, but rather makes up for the slow speed of the computation by simultaneously computing many processes and alternatives. This sort of increase in computational power far outweighs the raw computing speed; an advantage of the computer vs. the brain.

[c] A column consists of many neurons grouped together.

h. Notion of simple and complex cells.

   i. Simple cells perform a direct role and have more competitive effect among the cells.
   ii. Complex cells perform first order and second order combination, such as performing object category.

i. Mirror neurons. Imitation is one of the key principles in human learning. The ability to imitate is said to be due to the existence of mirror neurons. The principle of mirror neurons and how this arises due to the higher cognition can be learnt and modeled after in the smart system.
j. Dopamine. Dopamine is found to be related to how we learn via rewards. Current computational models that use reinforcement learning hard code rewards (extrinsic rewards). Perhaps the study of dopamine can give insights into developing a system that uses intrinsic rewards (a system that self-assigns reward values to various states).
k. Receptor and second messenger. Receptors are proteins within the cell nucleus that bind to a specific molecule such as neurotransmitters, hormones, or other chemical substances. Once a receptor is activated, it will in turn activate a second messenger, which will cause a cascade of actions in a cell. This behavior can be modeled for the computational process. For example, when an event is captured and activated (like a receptor waiting for activation), it would lead to a cascade of actions occurring.

2. At the network and system levels where neurons are arranged in groups/columns[d] and functional regions[e]:

   a. Hierarchical organization.

---

[d] Groups and columns usually only apply to relatively low-level sensory regions.

[e] A functional region could be a general region such as the associative cortex or a more specific functional region such as the Fusiform Face Area (FFA). It is noted that some functions and current ways of regionalizing the brain could be a debatable issue.

   i. Different layers of neural networks perform different tasks.
   ii. The layer complexity in performing tasks increases from the low-level sensory to high-level cortex.
   iii. Receptive field. The idea of a small receptive field in a simple cell to a large receptive field in a complex cell.

b. Regionalized into functions.

   i. Neurons of similar function are grouped together.
   ii. Neurons from one region can communicate with several other regions.
   iii. Some regions have very localized specific functions. For example, the Fusiform Face Area (FFA) is for the storing of faces and is used for face detection.

c. Feedback and feedforward process. The architecture can inherit such feedback and feedforward connections for information passing. For example, feedforward is good for fast processing and quick recognition and reaction.
d. Complementary role. Multiple subsystems working in complementary roles. For example, the dorsal and ventral streams of the visual pathway work complementarily to enable us to recognize and grasp an object.
e. Competitive role. Within the same group of neurons, there are competitive effects.
f. Cooperative role. Different regions and areas of the brain work cooperatively together to result in cognitive intelligence.
g. Multi-modality. Multiple sensor integration such as integrating auditory and visual operating together can be done.
h. Auto-calibrates and adapt to new demands or environment. Dynamically select good sets of parameters to work with.
i. The brain appears to leave data sources fragmented. Why? How can its data representation be created and modified?
j. Specific mechanism. The brain appears to have specific mechanisms that enable effective top-down and bottom-up

control. This process may be adapted in our intelligent system design. For example a top down control process based on "prior events" is specifically targeted to the relevant feature-specific area to boost conflict resolution in the brain.

k. Cognitive connections. The brain's cortico-cortical connections are interesting to mimic after. For example, information from an external stimulus reaches the amygdala (for generating emotional response) in two different ways. One is a short and fast, direct route from the thalamus; the other is a long and slow, indirect route by way of the cortex. The short and direct route lets us start preparing for a potential danger before we even know what it is. For example: let us suppose, while you are walking in a forest you suddenly see a long, narrow shape coiled at your feet. This snake-like shape leads you to jump up due to the short, direct route. But this same visual stimulus, after passing through the thalamus, will also be relayed to your cortex (where higher reasoning process) — a longer, indirect route. A few fraction of a second later, the same shape you thought to be a snake was really just a discarded piece of garden hose. This type of direct and indirect connections can improve the cognitive architecture by considering multiple decision paths and results, as more information is received and processed.

3. At the organismal level:

   a. Global stability and robustness.

      i. Shunting inhibition to prevent saturation. The brain does not suffer unexpected saturation and hence failed to work in its operation of the massively parallel neural networks.
      ii. Learn in a stable manner. The brain learns quickly and stably without catastrophically forgetting its past knowledge. Professor Stephen Grossberg calls this the stability-plasticity dilemma.

b. Predictive ability. Our brain has the ability to predict. For example, when we are juggling balls, our brain is able to predict the balls' movements and extrapolate the motor motion of our hands in parallel to catch the falling balls. In the juggling ball example, you look at the top level of the ball movement and you project where it is going to fall and you shift your hand to catch it. You do not look at the ball's complete movement (for if you do you will not have time to catch the ball) and you do not look at the bottom where the ball will fall and move your hand (again you will not have enough time to catch it, if you do so). Does our brain use the same machinery to perform all other predictions?

c. Pattern recognition. Our brain is very good at recognizing patterns and performs inference or prediction. Pattern recognition plays a key role in our abilities to solve problems. The brain is able to take in multiple features to recognize a pattern or multiple patterns.

d. Invariant object recognition capability and the ability to generalize.

   i. When we look at an apple from different angles, our retinas receive different views of the apple but we can still recognize that it is the same apple.
   ii. When we view several pictures of different dogs, we are able to generalize it into a single concept of "dog", and when we see a picture of another dog that we have not seen before, we can still recognize it as a dog.

e. The learning principle. We learn because the brain is plastic or it has neural plasticity. This may provide idea to generate

   i. "Plastic" program, i.e. program that is flexible and changes dynamically with environmental input.
   ii. "Dynamic" pointer, i.e. pointer that can dynamically link its association to involve other pointer or programs.

f. Different regions in the brain seem to adapt different synaptic plasticity. For example:

   i. Basal Ganglia — reinforcement learning.
   ii. Cerebellum — motor learning/error-based learning.
   iii. Hippocampus and cerebral cortex — Hebbian in general.

g. Automate most processes once it is learnt.

   i. Once learnt, the system does not need to be preprocessed. It sort of goes into a memory recall mode where stages are performed automatically. An example of this is learning to drive a car. Once learnt, the procedure for driving is automatic and the attention goes to the dynamic environment on the road.

4. Other design principles are:

   a. Attention mechanism. Where should I look and where should I focus?

      i. Interesting activities, such as peculiar, suspicious, and funny stuff lead us to pay special attention.
      ii. Statistically unusual, unexpected, and abnormal activities also catch our attention mechanism.

   b. Both localized and distributed. Some functions are pretty localized, such as recognizing faces (in FFA areas), which could be specific to a set of neurons performing specific tasks. Others appear more distributed, i.e. involve many sets of neurons forming different areas to perform a task such as navigation.
   c. Memory.

      i. Associate related activity and post-event information. In some sense, we can use a simple association technique to associate activities and related events.
      ii. Consolidation and formation. Since the memory of an event is consolidated from bits and pieces, Daniel

Schacter wrote that "*Memories for individual events resemble jigsaw puzzles that are assembled from many pieces*," and suggested that all persons in recollection, normally "*knit together the relevant fragments and feelings into a coherent narrative or story.*"

  - Consolidation during sleep. Some form of "after dark" can be emulated to perform computational memory reorganization like the brain does during sleep.
  - Memory formation for encoding, storing, and retrieving.

d. 3-D computing. As the brain neuron connections are 3-D, the layer connections of the artificial neural network design can be 3-D (multiple layers of 2-D stacked one after another in all directions).

e. Open-ended-open-minded design.[f] This refers to the brain's ability to process any kind of information and rewire its architecture. It is observed that the more complex cortical regions in the brain, such as the association cortex, may have relatively less pre-wired knowledge. This may be one aspect that enables the brain to process any kind of information.

When computers are one level smarter but have yet to reach human intelligence, we can use computers in a dual augmentation strategy to assist human intelligence. The next section will discuss this dual augmentation strategy.

Dual Augmentation Strategy

Currently, computers outperform humans in structured tasks. Humans are dynamic and outperform computers in all other

[f] Despite the pre-wiring, the brain is left with an open-ended-open-mind design, i.e. the brain can learn new functionalities.

non-structured tasks. Computers can be used to augment human intelligence, and likewise, humans can be used to augment computers, which in turn can assist other humans. How does this work?

Using humans to solve problems that computers cannot solve.

- For example, in the Mozes Mob case: Mozes Mob (mozes.com/mob.php) is a free, cell phone-based service designed to answer questions. To see how it works, text message a question that a human could answer easily but that a computer program would have a hard time figuring out (such as "Is the weather nicer in Miami or Buffalo?" or "What was Carlton Fisk's most famous home run?") to 66937. Behind the scenes, your question is sent to a swarm of volunteers. One of them answers, and their response is bounced back to your cell phone, usually in just a few seconds.
- Another example is the Google image labeler which (images.google.com/imagelabeler) is an addictive online game that takes advantage of the fact that it's very easy for a human to recognize the subject matter of an image (e.g. "That's a puppy!" or "That's two airplanes and a bird!") but virtually impossible for a computer. The game teams up pairs of strangers to assign keywords to images as quickly as possible. The more images you can label in 90 seconds, the more points you get. Meanwhile, Google gets hundreds of people to label their images with keywords, something that would be impossible with just computer analysis.

These two examples use human society to augment the computer. The augmented computer then appears intelligent and aids other humans. Although the current computer appears more like a relay tool, in the long run, when it has more intelligence, the dual augmentation strategy can be made more sophisticated and humans can get the best of the computational power and its intelligence.

Computational Power

The computational power of a machine can continue to evolve with its designer, while the human brain design is already fixed or some believe that it will evolve at a much slower pace than the computer. Hence, humans can leverage on the computational machine since the machine:

1. Can be designed to remember all facts and knowledge. Humans can forget.
2. Does not suffer fatigue like humans, and hence can be used to complement humans.
3. Does not suffer from depression. It can be designed with emotions that help in its communication with humans. Negative emotions such as depression need not be built into the system.
4. Size need not be constrained to brain-size. Hence, the computer can have more memory systems built-in than current brain memory.

Conclusion

The human quest for building intelligent systems to assist in their work is inevitable and necessary. This chapter discussed some key design principles we can learn from the biological brain to build intelligent systems.

So far, many intelligent or smart systems developed are for specific applications such as intelligent transportation systems, self-diagnostic systems, "smart" home appliances like the autonomous vacuum cleaner and other military systems. In time to come, we hope these intelligent systems can be enhanced in terms of the architectural, functional, power consumption and algorithmic design by taking inspiration from the brain, or as US DARPA SyNAPSE project initiative in seeking to pick the most useful elements of the brain, to achieve higher-level cognition and to perform more general tasks.

"*The closer we can get to human-like cognition, the better.*" — David Gelernter (Gelernter, 1994) (The Free press).

Chapter 14

CONCLUSIONS — THE MIND THAT MATTERS

"... *the mind is what the brain does.*" — National Geographic, "Beyond the Brain", March 2005 article by James Shreeve and Cary Wolinsky.

In a short span of 50 years, computer size has dropped from one of room size to today's palm-size, and processing power has increased from a few hundred instructions per second to three giga hertz or more, i.e. three billion instructions per second, and with multiple processors working together. Technology has enabled us to communicate and meet people from different time zones and places seamlessly. 100 years ago, this would have been unimaginable. Why has this happened within such a short span of time? This is especially remarkable given that humans have been in existence for thousands of years with the same brain capability.

"*Throughout the recorded and archeological history of humankind, the way of life that the majorities have followed has been dramatically shifted by technological revolutions. Technological revolutions are happening more and more often. The dawn of the most recent of these revolutions, the digital revolution, is within living memory of many of us alive today. The battles and restructuring of our world brought on by the digital revolution is raging around us at its full fury today. It is*

> *shifting wealth and power, and restructuring our cities and lifestyles.*" — Rodney Brooks (Brooks, 2003).

Using these advanced digital technologies, many researchers are embarking on the journey to understand the brain for many reasons. Till today, the brain remains a wonder to scientists and an area that can generate many research topics. Interesting new results and new hypotheses on the brain machinery and how it gives rise to a mind will continue to overwhelm us. It is no surprise that every now and then we will hear or read new discoveries or new theories of the brain.

One of the reasons for understanding the brain, as discussed in this book, is to build computationally intelligent systems that have some human-like intelligence. However, some scientists would desire to have the full or identical human-like intelligence. If the quest to meet the complete human intelligence can be achieved one day, then the machine theoretically will have a mind of its own. What do we then say of this mind that arises out of a computational box?

The brain is living machinery and it gives rise to a mind. Ultimately, it is the mind that matters and how we use the mind. This chapter will discuss the current thinking on the theory of the mind, and what is the meaning of having a mind. I will also relate the theory of my mind.

What are Theories of the Mind?

What is the mind? Based on:

- Webster's Dictionary, the "mind" means the part of an individual that feels, perceives, thinks, wills, and especially, reasons.
- Wikipedia.org/wiki/Mind, The term "the mind" is most commonly used to describe the higher functions of the human brain, particularly those of which humans are subjectively conscious, such as personality, thought, reason, memory, intelligence, and emotion. Although other species of animals share

some of these mental capacities, the term is usually used only in relation to humans and supernatural beings to which human-like qualities are ascribed, as in the expression, "the mind of God".

The phrase "theory of mind" (or TOM) is used in several ways and has different meanings, although they are in general related to one another, i.e. trying to understand the mind. TOM is interdisciplinary and used by philosophers, psychologists, and cognitive scientists. There are currently a few interrelated schools of research with regards to the "theory of mind". These are:

- Philosophy of mind. The study of what is the mind, the nature of the mind, the mental events, structure, functions, properties, consciousness, processes, and their relationship to the physical body (commonly known as mind-body problem, mind and matter, also related to the mind-brain identity). This attempt to address issues such as whether a non-material mind (i.e. mental such as pain, desire, belief, purpose, etc.) can influence a material body (physical body) and vice-versa. For example, the mental state "desire for a cup of coffee" will cause the body to move in a specific manner and in a specific direction to obtain what they want. In other words, our perceptual experiences depend on stimuli that arrive at our various sensory organs from the external world, and these stimuli cause changes in our mental states. These mental states cause us to have feelings of sensation, both pleasant and unpleasant. (Gopnik, 2003) (Carey, 1985) (Wellman and Gelman, 1992) (Carruther, 1996).
    - Cartesian dualism.[a] (Same as mind-body dualism.) The belief is that there exist two distinct entities, namely body and soul,

[a] Theory of theory. Some scientists prefer this approach via Cartesianism and neo-Cartesianism as the understanding of mentalistic notions (such as belief, desire, perception, intention, etc.) on how people behave in relation to the structure and function of the mind and cognitive abilities (Carruther, 1996).

and that there is a clear distinction between the physical body (including the brain) and the non-physical mind. These two entities interact with each other causally, though it is not known how.

  - Mind-brain identity. The identity theory of mind-brain holds that the mental states and processes of the mind are identical to brain states and the process of the brain, i.e. the mental state "A" is nothing more than the brain state "B". For example, the mental state, "desire for a cup of coffee", would be a brain state, the firing of certain neurons in certain brain regions (Smith, 1956).

- Reading people's minds. Understanding and interpreting other people's actions and behaviors, i.e. what is their belief, desire, and intention (when people think about other people's thoughts). Other names include mind reader and mind reading (Davidson, 1984) (Dennett, 1987). For example, when you see a child crying while a fire alarm is off in his home, you can infer that the child is afraid. Or if you see a young couple waving happily and saying goodbye at a train station to an older couple, and the older woman is crying and the old man says something and hugs her, you can probably guess that the old man is saying some comforting words to his wife (the older woman) and the young couple is probably newly married and going on a honeymoon. Note that the scope of TOM research is focused not on psychic power[b] but using the understanding of one's self to understand another person and to socialize effectively. Many mind-reading theories have been studied in parallel with autistic individuals. This is because autistic individuals seem to have an impairment of their "mind-reading" ability. And the study of TOM may help autistic children during their early development process.

---

[b] Psychic power here refers to the unusual mental powers that cannot be explained or extraordinary abilities to tell what people are thinking and able to see into the future.

"*You never really understand a person until you consider things from his point of view... until you climb into his skin and walk around in it.*" — To Kill a Mockingbird by Harper Lee (Helper, 2003).

- Computational theory of mind. This approach is used by scientists, particularly interested in artificial intelligence or cognitive intelligence, to model the mind based on understanding the nature of the mind. This includes:
  - Structure and process of the mind.
  - Functional theory of mind, i.e. how the mind works and its various possible functions.
  - Semantic properties of the mental state and how this is represented in the mind.

Next, computational techniques are used to model how the mind works and to understand the working mechanism of the mind.[c] See (Horst, 2005) for the computational theory of mind. Issues on whether the theory of mind or the mind itself and the cognitive process can be modeled fully by computational techniques are discussed and debated. Can a human build a machine that has a mind of its own — like the human mind? The reader interested in more in-depth discussion may refer to (Searle, 1990; 1991; Edelman, 1989; Gibson, 1979), which raise several questions regarding this matter; and (Chalmers, 1993) and (Kurweil, 1999), where it is thought that it is possible to fully model the mind by the computational approach.

[c] The study on brain-inspired computing is related to the computational theory of mind. Brain-inspired computing takes from the biological perspective of the brain and computational theory of mind takes from the behavior perspective of the mind. They are related to the same quest in building intelligent systems that have human-like intelligence.

The Mind that Matters

What is the conclusion of the matter? "*What is mind? No matter! What is matter? Never Mind.*" — C.U.M. Smith.

"*The advance of modern neurophysiology has both sharpened the Cartesian dilemma and at the same time tended to obscure it. For few of us realize this scandal in the depth of our culture: this schizophrenia. For, on the one hand we feel bound to assert that minds do in fact act upon bodies, and on the other that they do not so act.*" — (Smith, 1970).

We have not come to a clear conclusion of what is the mind. What is the mental state and what is the mental representation of the mind? How the physical brain gives rise to a mind remains largely a mystery. The human brain is so complex, it gives rise to a mind that can investigate the laws of the universe, question its existence, and ask about the nature of God.[d]

Professor Stephen Hawking, in his study of the universe, raised the issue that if a complete theory can be found to explain the universe then we would know the mind of God. In his own words, "*... if we do discover a complete theory, it should in time be understandable in broad principle by everyone, not just a few scientists. Then we shall all, philosophers, scientists, and just ordinary people, be able to take part in the discussion of the question of why it is that we and the universe exist. If we find the answer to that, it would be the ultimate triumph of human reason — for then we would know the mind of God.*" — A Brief History of Time by Steven Hawking (p. 175).

"*Up to now, most scientists have been too occupied with the development of new theories that describe what the universe is to ask the question why.*" — Steven Hawking.

---

[d] Seeking to understand the mind of God.

Are we living in the same situation where scientists are too occupied with the uncovering of how the brain works to ask the question of why?

An article from The New York Times on "Who's minding the mind?" (Carey, 2007) highlighted the recent study by Professor John A. Bargh[e] and his co-researcher Lawrence Williams: "*When it comes to our behavior from moment to moment, the big question is "What to do next?" Well, we are finding that we have these unconscious behavioral guidance systems that are continually furnishing suggestions through the day that what to do next and the brain is considering and often acting on those, all before conscious awareness. Sometimes those goals are in line with our conscious intentions and purposes, and sometimes they are not.*"

The same article concluded that "*...the new research on priming makes it clear that we are not alone in our own consciousness. We have company, an invisible partner who has strong reactions about the world that don't always agree with our own, but whose instincts, these studies clearly show, are at least as likely to be helpful, and attentive to others, as they are to be disruptive.*" The New York Times, July 31, 2007 (Who's Minding the Mind?) (Carey, 2007).

It is amazing that we can see, hear, and think of our world, the universe, and read and imagine whatever is before and after us. Anything within this physical space where we, as human beings, are bound, we can imagine and attempt to understand. The mind is what matters; how we think has become what the brain creates, giving us the understanding and ability to appreciate love, joy, pain, and suffering in the world we live in.

> "*Then He opened their minds so they could understand ...*" — Luke 24:45.

---

[e] A professor of psychology at Yale University.

GLOSSARY

Acetylcholine: A neurotransmitter released at the neuromuscular junction. Also released from certain synapses in the brain, where it can have either neurotransmitter or neuromodulatory effects, and from parasympathetic nervous system neurons.

Action potential: The transient of electrical signals that travel down axons carrying the output information of neurons. Basically, it is simply an electrical current that travels down an axon of a neuron.

Agent: See multi-agent systems.

Amino acid: The molecules, which when strung together, form proteins.

Amygdala: Located in the limbic system. Center of emotional chemical secretion and plays an importance role for emotional behaviors such as fear, joy, and aggressive behavior.

Anterograde amnesia: Memory loss due to an inability to transfer information from short-term memory to long-term memory. Damage to hippocampus can cause anterograde amnesia. (See also retrograde amnesia).

Anterior commissure: A forebrain commissure that provides communication between structures within the temporal lobes.

Artificial intelligence (AI): The study of human cognitive abilities (intelligence) using models that are implemented as computer programs or hardware. AI used to be distinct from artificial neural networks, however, this dividing line is getting blur.

Artificial neural networks (ANN): ANN is a computer program whose structure resembles the brain's neural network connection, i.e. an individual neuron is modeled and many of these neurons are interconnected in networks. They are sometimes known as connectionist models. The goal of ANN is to simulate or emulate the performance of tasks of which animals or humans are capable.

Association cortex: Area in the cerebral cortex that concerns higher level of processing.

Autoreceptor: A receptor molecule located in the presynaptic neuron's axon terminal. It is thought to play a role in providing feedback to the presynaptic neuron and also in modulating synaptic activity.

Axon: Part of the neuron that carries the neural impulse away from the soma toward the target of the neuron. This part of the neuron carries action potentials.

Axon terminal: The ending of the axon that connects to the neural target. It contains the neural transmitters to be released.

Autonomic nervous system: That part of the nervous system that regulates our internal organs. Much of the regulation is involuntary and mediated by two opposing subdivisions, the sympathetic and parasympathetic systems.

Axon: It carries information away from the soma to terminal buttons where messages pass to other neurons, muscles, or glands.

Central nervous system (CNS): The CNS consists of the brain and spinal cord.

Cerebral cortex: The largest part of the brain cortex, highly infolded in humans. The cerebral cortex is divided into two hemispheres, which are further subdivided into four lobes: frontal, parietal, occipital, and temporal.

Chloride (Cl–): A negatively charged ion primarily involved in the inhibition of neurons.

Cingulate gyrus: This is a gyrus in the medial part of the brain. The cingulate gyrus receives inputs from the anterior nucleus of the thalamus and the neocortex as well as from somatosensory areas of the cerebral cortex. It functions as an integral part of the limbic system.

Cognitive: Related to the mind and brain.

Cognition: Related to the operation of the mind process by which awareness of objects, thought, and perception takes place.

Cognitive neuroscience: A field of scientific study on biological mechanisms underlying cognition, with a specific focus on the neural substrates of mental processes and their behavioral manifestations. It addresses questions of how psychological and cognitive functions are produced by the neural circuitry. Cognitive neuroscience is a branch of both psychology and neuroscience. It overlaps with many disciplines such as cognitive studies, psychology, psychobiology, neurobiology, physics, psychiatry, neurology, and mathematics.

Cognitive psychology: A field of study on human cognitive abilities such as decision making and reasoning abilities.

Cognitive science: An interdisciplinary study of the mind and intelligence, embracing philosophy, psychology, artificial intelligence, neuroscience, linguistics, and anthropology. Its intellectual origins are based in the mid-1950s when researchers in several fields began to develop theories of mind based on complex representations and computational procedures. Its organizational origins are in the mid-1970s when the Cognitive Science Society was formed and the journal, Cognitive Science, began. The words "cognitive science" were coined by Christopher Longuet-Higgins in 1973.

Commissure: Any fiber tract that connects structures on the two sides of the central nervous system.

Computational neuroscience: A field of study of the nervous system that utilizes tools from mathematics and computer science.

Corpus callosum: A thick band of axons, found in the middle of the brain, that carries information from one hemisphere of the brain to the other hemisphere. It is the largest commissure. It provides communication across the two hemispheres.

Critical period: The period of time during development when an animal is particularly sensitive to environment conditions.

CT scan: A series of x-ray photographs taken from different angles and combined by computer into a composite representation of a slice through the brain (CATscan). CT scans are useful in the study of brain structures.

Current: Electrical energy flowing — that is, the kinetic energy form of electricity.

Dendrites: These bushy branch-like structures are a neuron's message receivers. Dendrites receive impulses from other neurons (via synapses) or from sensory organs, and bring them toward the cell body, i.e. soma.

Diffusion: Diffusion is the effect that the random movement of molecules has, which causes the movement of molecules from a region of high concentration to low concentration. Diffusion (the change in the relative levels of concentration) will stop when the concentration of the substance in solution is equal at all points.

Dopamine: A neurotransmitter released from brain synapses. Dopamine is associated with Parkinson's disease and schizophrenia. Dopamine is also a neurohormone released by the hypothalamus.

Fovea: The central region of the eye that mediates high-acuity vision.

Free recall: The production of material from memory without the aid of specific cues.

GABA (gamma-aminobutyric acid): It is one of the most common inhibitory neurotransmitters. GABA binds to a receptor with a channel that selectively passes chloride ions. Upon entry, these ions increase the negative potential of the cell and thereby inhibit firing.

Ganglion cells: The third-order cells in the retina whose axons form the optic nerve.

Gray matter: Those regions of the brain and spinal cord where neuronal cell bodies and dendrites are abundant.

Glutamate: A major excitatory neurotransmitter involved in memory. Oversupply can over stimulate brain, producing migraines or seizures (which is why some people avoid MSG, monosodium glutamate).

Hippocampus: A region of the brain found under the temporal lobes that play an important role in memory formation (particularly long-term memory) and the representation of space (e.g. place cells).

Hypothalamus: A forebrain region that regulates autonomic, endocrine, and visceral functions. Hence, it is concerned with basic acts and drives, such as drinking and sexual activity.

Huntington's disease: An inherited disease of the basal ganglia that causes movement dysfunction.

Inferior temporal cortex (IT): The IT is necessary for visual pattern discrimination, and receives information from the visual cortex.

Ion: An electrically charged particle. Ions can be positive (e.g. Na+) or negative (e.g. Cl–). Like charges repel or push each other away, and opposite charges attract or draw each other.

Limbic system: The limbic system is involved in emotion formation and processing, learning and memory. The system consists of the thalamus, hypothalamus, amygdala, olfactory bulb, hippocampus, and basal ganglia. (Note that the basal ganglia is part of the limbic system. The basal ganglia lie beside the limbic system and is tightly connected to the limbic system and the cortex above it).

Mirror neurons: Mirror neuron is a neuron that fires or is activated when an animal performs an action and when the animal observes the same action performed by another animal, i.e. the neuron "mirrors" the behavior of another animal as though the observer were itself performing the action. Mirror neurons have been interpreted as the mechanism by which we simulate or imitate others in order to better understand them (Gallese and Goldman, 1998) (Iacoboni *et al.*, 1999).

Myelin: A sheath of fatty tissue that covers most axons on the nervous system. It serves to speed conduction, and hence, increase the speed of transmission of neural impulses.

Moore's law: It states that transistor density on integrated circuits doubles about every two years.

Multi-agent system: In computer science, a multi-agent system (MAS) is a system composed of several autonomous agents, collectively capable of reaching goals that are difficult to achieve by an individual agent or monolithic system.

Neural plasticity: A term from neuroscience indicating changes in the brain that alter brain function so as to constitute learning. It is assumed that with the structure of the brain fixed, responding will not change. Learning is presumed to be due to changes in the neural structure, and hence, neural plasticity.

Neurogenesis: The process by which neurons are created or the birth of neurons. Most active during pre-natal development, neurogenesis is responsible for populating the growing brain. Growth of new neurons.

Neurophysiology: A study of nervous system function. Neurophysiology is part of physiology and is closely related to psychology, neurobiology, neuroanatomy, cognitive science, and other brain sciences.

Neuroscience: A field of science for studying the cellular level of neurons in the nervous system.

Neuron: The nerve cell, the basic building block of the nervous system.

Neural networks: A collection of abstracted neurons connected to each other through weighted connections ("synapses"). See also artificial neural networks (ANN).

Neurotransmitters/neuromodulators: These are chemicals substances released from a neuron at a synapse that are used to relay, amplify, and modulate electrical signals between a neuron and other neurons. Neurotransmitters convey information between two neurons, but neuromodulators convey information to a region of neurons (or group of neurons).

Nucleus: The structure in the soma of the neuron that contains the chromosomes.

Optic nerve: The portion of the axons that originate in the ganglion cell layer of the retina that travels from the retina to the optic chiasm.

Occipital lobes: Involved in processing visual information.

Orbitofrontal cortex: An area found in the lower part of the frontal lobes, important for the expression of emotional behaviors.

Parietal lobes: Top and rear side of the brain, involved in processing somato-sensory information (touch).

Potential energy: Energy that is available to do work.

Parkinson's disease: A disease of the motor system caused by a deficiency of dopamine in the basal ganglia. Patients with the disease typically develop a tremor and have difficulty initiating movements.

Peptide: A small protein.

Plasticity: The brain's capacity for modification, as evident in brain reorganization following damage (especially in children) and in experiments on the effects of experience on brain development. The brain can rewire itself with new synapses or select new uses for its prewired circuits.

Postsynaptic Neuron: The neuron that receives the neurotransmitters released by the presynaptic neuron. It is in most synapses a dendrite or soma.

Receptor: A molecule in the postsynaptic neuron's membrane to which the neurotransmitters bind, which then directly or indirectly alter the membrane potential by opening or closing ion channels.

Potassium (K+): A positively charged ion primarily involved in establishing the resting potential.

Primary motor area: The region of the cerebral cortex where fine movements are initiated. Found in the frontal lobes adjacent to the central sulcus.

Primary sensory area: Regions where sensory information is first processed in the cerebral cortex.

Psychophysics: A subdiscipline of psychology dealing with the relationship between physcial stimuli and the observer or subject's point of view.

Purkinje cell: A large neuron found in the cerebellum.

Pyramidal cell: A prominent neuron found in all areas of the cerebral cortex.

Receptive field: The receptive field of a sensory neuron is a region of space in which the presence of a stimulus will alter the firing of that neuron. The concept of receptive fields can be extended to further up the neural system (information from wiki).

Retina: Interior lining of the back of the eye containing photoreceptors.

Retrograde amnesia: Memory loss where one cannot recall events that occurred before the onset of amnesia. The memory loss may only affect some types of memory. For example, a pianist with retrograde amnesia may not remember how to play the piano but can remember how the instrument looks.

Saccade: A saccade is a fast movement (or jerky movement) of both eyes in the same direction. We make about three saccades in one second, each lasting between 20 and 200 microseconds.

Second messenger: A small molecule synthesized in a cell in response to a neuromodulator (the first messenger).

Singularity: A point in space-time at which the space-time curvature becomes infinite (extracted from "A Brief History of Time" by Stephen Hawking).

Soma: The cell body where the nucleus is found. Most of the protein production and energy storage is performed by soma for sustaining the life of a cell.

Sodium (Na+): A positively charged ion involved in the generation of action potentials and in the excitation of neurons and sensory cells.

Somatosensory: Pertaining to sensory information coming from the skin and deeper tissues of the limbs and trunk, such as touch, pressure, temperature, and pain.

Split-brain: This is a term to describe the result when the corpus callosum connecting the two halves of the brain is severed to some degree.

Stability-plasticity dilemma: A term introduced by Professor Stephen Grossberg. It is concerned with the fact that our brains can rapidly learn enormous amounts of information throughout life, without just as rapidly forgetting what is already known. The brains are plastic and can rapidly learn new experiences without losing the stability that prevents catastrophic forgetting (Grossberg, 1999).

Substantial nigra: This is a concentration of neurons in the ventral portion of the midbrain that uses dopamine as its neurotransmitter and is involved in both motor function and emotion. Its dysfunction is implicated in Parkinson's disease.

Sulcus: A groove or deep infolding on the surface of the brain.

Superior colliculus (SC): The two pair of bumps on the posterior surface of the midbrain, just below the thalamus. SC plays a role in vision and is said to be responsible for the generation of saccadic eye movement and eye-head coordination.

Synaptic Vesicle: A membrane structure in the axon terminal that contains neurotransmitters until they are released.

Thalamus: Innermost part of the forebrain, it relays sensory information to the cerebral cortex.

Uncertainty principle: One can never be exactly sure of both the position and velocity of a particle; the more accurately one knows the one, the less accurately one can know the other (adapted from "A Brief History of Time" by Stephen Hawking).

White matter: Regions of the brain and spinal cord where there are abundant myelinated axons. The myelin gives the tissues its whitish appearance.

System neuroscience: This is a sub-disciple of neuroscience, which studies the neural circuit function, most commonly in awake, behaving intact organisms. The system neuroscience research area is concerned with how nerve cells behave when connected together to form neural networks for a common function such as vision.

NOTES

Chapter 1

1. Phrenology is a theory developed by German physician Franz Joseph Gall in the eighteenth century. He claimed that by the basis of the shape of the head, one can determine character, personality traits, and criminality. Gall's map of the brain is disputed by modern neuroscience. However, Gall's theory that the cerebral cortex does not act as one organ but as a collection of specialized regions, and that the brain is the basis of all behavior remain the truth as of today.
2. The average human male brain has two billion more neurons than the average female brain. However, in women, the neuron connection is stronger than in men. This may explain why males have better spatial reasoning and females better language ability. Males have an average of 299,052 neurons per square mm on the surface of a brain slice about 3 mm thick. Females have an average of 264,892. (Rabinowicz *et al.*, 2002).
3. Most complex mental functions such as learning and memory don't reside in any one place but are distributed throughout the brain.
4. Dendritic spine.
   The figure shows the dendrite spine and spine head. Information and picture from http://en.wikipedia.orgg/wiki/Dentritic_spine.

   The dentritic spine is a small membranous extrusion that protrudes from a dendrite and forms half of a synapse. Spines are typically connected to the parent dendrite through a thin spine neck as in the figure. Dendritic spines are found on the dendrites of most principal neurons in the brain. Spines come in a variety of shapes and have been categorized accordingly, e.g. mushroom spines, thin

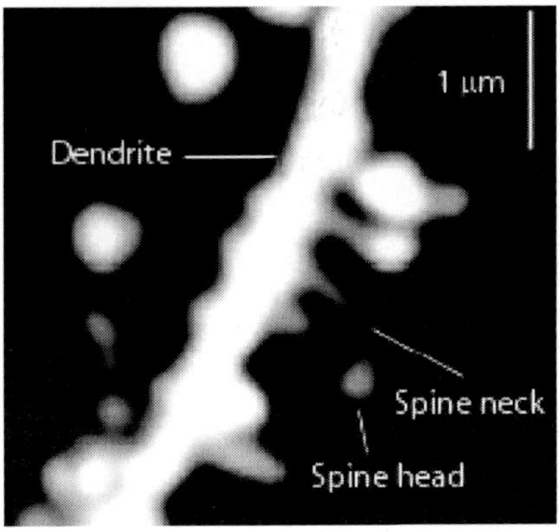

spines, and stubby spines. Electron microscopy studies have shown that there is a continuum of shapes between these categories. There is some evidence that differently shaped spines reflect different developmental stages and also strengths of a synapse. Using two-photon laser scanning microscopy and confocal microscopy, it has been shown that the volume of spines can change depending on the types of stimuli that are presented to a synapse.

The detailed brain anatomy, explanation of each part, and their respective functions can be found in the following websites:

- http://www.waiting.com/brainfunction.html
- http://thalamus.wustl.edu/course/cerebell.html
- http://www.stanford.edu/group/hopes/basics/braintut/ab0.html
- http://www.lib.uiowa.edu/hardin/md/brainpictures.html.

Chapter 2

1. More information on neurons and synapses can be found in the following websites:

   http://faculty.washington.edu/chudler/neurok.html
   http://www.cerebromente.org.br/
   http://www.biologymad.com/NervousSystem/

2. Hodgkin–Huxley's cell membrane equation. (Hodgkin and Huxley, 1952) (Nelson, 2004):

$$C\,dv/dt = (V^{+} - V)g^{+} + (V^{-} - V)g^{-} + (V^{p} - V)g^{p}$$

g^{+} is the excitatory term
g^{-} is the inhibitory term
g^{p} is the passive decay term

3. *Simple spiking model.* The simple spiking model proposed by Izhikevich claimed to reduce many biophysically accurate Hodgkin–Huxley type neuronal models to a 2-D system of ordinary differential equations. The details of the ordinary differential equations are as follows:

$$V' = 0.04\,v2 + 5v + 140 - u + I$$
$$U' = a(bv - u)$$

With the auxiliary after-spike resetting,

If $v \geq 30$ mV, then

Here, v and u are dimensionless variables, and a, b, c, and d are dimensionless parameters where t is the time. The variable v represents the membrane potential of the neuron and u represents a membrane recovery variable of Na^{+} ionic currents, and it provides negative feedback to v. After the spike reaches its apex (+30 mV), the membrane voltage and the recovery variable are reset according to the equation 3. Synaptic currents or injected dc-currents are delivered via the variable I.

The part $0.04\,v2 + 5v + 140$ was obtained by fitting the spike initiation dynamics of a cortical neuron so that the membrane potential v has *mV* scale and the time t has *ms* scale. The resting potential in the model is between –70 and –60 mV depending on the value of b. As with most real neurons, the model does not have a fixed threshold; depending on the history of the membrane potential prior to the spike, the threshold potential can be as low as –55 mV or as high as –40 mV (Izhikevich, 2003).

4. *Major neurotransmitters and their functions.*

   a. *Acetylcholine* (ACh): Enables muscle action, learning, and memory. Deficiency implicated in Alzheimer's disease.

b. *Dopamine*: Influences movement, learning, attention, and emotion. Excess dopamine receptor activity is linked to schizophrenia; starved of dopamine, the brain produces the tremors and decreased mobility of Parkinson's disease.
c. *Serotonin*: Affects mood, hunger, sleep, and arousal. Undersupply linked to depression; Prozac and some other antidepressant drugs raise serotonin levels.
d. *Norepinephrine*: Helps control alertness and arousal. Undersupply can depress mood.
e. GABA (gamma-aminobutyric acid) is one of the most common inhibitory neurotransmitters. GABA binds to a receptor with a channel that selectively passes chloride ions. Upon entry, these ions increase the negative potential of the cell and thereby inhibit firing.
f. *Glutamate*: A major excitatory neurotransmitter involved in memory. Oversupply can over stimulate the brain, producing migraines or seizures (which is why some people avoid MSG, monosodium glutamate).

5. *The nervous system* (NS) is an electrochemical communication system, consisting of all the nerve cells of the peripheral and central nervous systems.

   a. Central NS (CNS) consists of the brain and spinal cord.
   b. Peripheral NS (PNS) consists of the sensory and motor neurons that connect the CNS to the rest of the body.

Chapter 3

1. *Motor cortex (MC)*. An area at the rear of the frontal lobes that controls voluntary movements; the left MC controls the right side of body and the right MC controls the left side of body; the larger area of the MC is devoted to areas of the body that require finer control (e.g. fingers vs. forearm); just thinking about movement produces electrical activity in the MC.
2. *Sensory cortex (SC)*. An area at the front of the parietal lobes that registers and processes body sensations; the left SC receives information from the right side of the body and the right SC receives information from the left side of the body; the larger areas of the SC are devoted

to the more sensitive areas (lips) of the body (e.g. lose a finger, then a greater SC area gets devoted to adjacent fingers; pianists have more auditory cortex, deaf people more visual cortex).

3. *Associative cortex (AA)*. The remaining 2/3 of the cortex involved in higher mental functions such as learning, remembering, thinking, and speaking, the AA integrates and acts on information received and processed by the sensory area; unlike the MC and SC, stimulating specific areas of the AC does not result in specific responses so functions cannot be neatly specified.
4. *Visual spectrum*. The range of electromagnetic radiation that is visible to us, 380 to 760 nanometers.
5. Perception of color determined by three dimensions:

   a. hue — determined by wavelength (what we think of as color).
   b. brightness — determined by intensity.
   c. saturation — relative purity of the light that is being perceived.

6. Three types of eye movements:

   a. vergence movement — cooperative, with both eyes kept fixed on a target.
   b. saccadic movements — abrupt shifts in gaze from one point to another, as when reading.
   c. pursuit movement — following the movement of an object.

7. Visual association cortex:

   a. ventral stream — ends with the inferior temporal cortex, involved with the perception of objects ("what").
   b. dorsal stream — ends with the posterior parietal cortex, involved with the perception of location, movement, and control of eye and hand movements.

Chapter 4

1. (Vincent *et al.*, 2006) presented data in support of the hypothesis that specific regions with the parietal cortex play a role in memory functions associated with the hippocampus formation. Furthermore, they show that those regions of the parietal cortex associated with the hippocampus formation are automically distinct from regions of

the parietal cortex traditionally associated with spatial attention and motor intention. More broadly, the hippocampus network, defined in their study, appears to be very similar if not identical to a group of areas posited to constitute a "default network". The fact that the posterior parietal components of this network are specifically modulated by recollection would be consistent with the putative self-referential nature of this network's other components.

2. The priming task involves:
   a. a showed list of words.
   b. asking for recall (explicit) or showing the first few letters and asking to say the first word that comes to mind (implicit).
   c. amnesic subjects' explicit memory was <50% of controls', but they performed equally on the implicit task.
3. Anterograde amnesia patients are not able to form new memories, or have problems learning new information.
4. Retrograde amnesia patients are not able to retrieve established memory. Or trouble remembering old information before the injury.
5. Amnesia patients are usually not affected by semantic memory although some kind of dementia can have an effect on semantic memory.
6. While many people tend to forget things, there are some individuals with superb memory abilities. An example is a subject known as "S", who employed mnemonic as a technique to enhance memory formation. This technique involves the conversion of a series of words into graphic images. These images are supposed to be vivid and stable, which means that the same image is brought up for the same word. These images are distributed along a visualized street, and one would recall information as he "walked" down this visualized street.

Chapter 5

1. *Is it possible to have a pure memory system without a learning process?* However, can we have no learning process and just a pure memory system as discussed in the chapter on memory systems? Take for example the patient HM, who has lost the ability to form new long-term memories due to the surgical removal of his medial temporal lobe to cure his epilepsy. HM lost approximately two-thirds of his hippocampal

formation, parahippocampal gyrus, and amygdale. After the surgery he suffered from severe anterograde amnesia and could not commit new events to long-term memory. However, he was able to perform meaningful conversations with his short-term working memory (Schaffhausen, 1997). Some may debate, using this example as evidence that the learning process may not be totally needed. However, even for this short-term working memory, there are changes in synapse, spike pattern, excitatory and inhibitory action, and the interaction with other synapses via neurotransmitters. These activities are discussed in Chapter 3 on synapses and neurons, that neuroscientists believe give rise to a learning process. However, there remain many puzzles, such as what this learning process is about. How can new information be captured in the memory? How is new memory formed given that there is relatively little new growth of neurons in the adult brain? How does the old memory get refreshed or remain in the system? What makes our learning so efficient and effective? Various researchers' findings will be presented and discussed here to give some insight into these questions.

2. *Long-Term Potentiation (LTP).* LTP is the earliest way in explaining the neural plasticity (Bliss and Lomo, 1973) (Kelso *et al.*, 1986). LTP demonstrates changing interactions between neurons and is said to underlie memory and learning. Two neurons connected by a synapse are identified either in vivo or in vitro. In LTP, correlated activity by both a pre-synaptic and post-synaptic neuron facilitates later transmission of signals from the former to the latter. In short, after both neurons are activated simultaneously, the conditional likelihood that the post-synaptic neuron fires, given that the pre-synaptic neuron fires, is raised.
3. *In-vitro reinforcement.* Stein and colleagues demonstrated a form of neural plasticity apparently distinct from long-term potentiation (LTP). They called it the in-vitro reinforcement (IVR). The basic result is that changes in neural activity are due to the infusion of the neurotransmitter, dopamine. A neuron that exhibits a characteristic multi-spike burst is monitored in-vitro. Whenever a burst is detected, dopamine is injected around the cell via a pipette. The burst rate is observed to increase. The basic notion is that initially random activity, when regularly followed by a biologically important event, will come to occur more often. (Stein, 1994) (Stein *et al.*, 1994) (Stein, 1997). Hence IVR involves increases and decreases in intrinsic burst

rates of individual neurons. IVR is presumed to produce learning by differentially reinforcing activity by those neurons involved in generating more adaptive behavior.

a. An artificial neural network with a computational model of learning based on a simulation of IVR has demonstrated that complex, independently-generated behaviors can be learned by a network of IVR-sensitive neurons. Thus, IVR may provide a key to understanding how complex, cognitive behaviors are learned.

4. *Hedonistic synapses* is another hypothesis proposed by Sebastian Seung from MIT, on how the neural plasticity work is different from IVR and LTP. Hedonistic means "reward seeking". In the case of neurons, Seung explained that the actions are the release of vesicles of neurotransmitters into the synaptic cleft and a reward corresponds to some chemical signal such as dopamine. Vesicle release from a presynaptic neuron is a stochastic process. When an action potential goes through the neuron, there is a probability that the neuron will release vesicles or fail to release them (hence, only two actions performed: either release the neurotransmitter or fails to release). Seung hypothesized that synapses are responding to a global reward signal by increasing their probabilities of vesicle release or failure, depending on which action immediately preceded the reward. Seung then modeld the hedonistic synapses learning by computing a stochastic approximation to the gradient of the average reward. Seung claims that HS is compatible with synaptic dynamics such as short-term facilitation and depression and with the intricacies of dendrite integration and action potential generation. A network of hedonistic synapses can be trained to perform a desired computation by administering reward appropriately, through numerical simulations of integrate-and-fire model neurons (Seung, 2003).
5. *LTP* involves variable sensitivity of synaptic connections between two neurons. IVR appears to involve variable sensitivity of a single neuron to its overall activity due to various causes. HS appears to involve a single neuron.
6. *Critical period.* Many scientists believe the notion of "use it or lose it" during the critical period of brain development. That is, if one misses the critical period or does not stimulus the brain during the

critical period, then one may lose the function or ability to learn, which is dependent on that plasticity during the critical period.

7. *Intrinsic reward mechanism.* In humans it has an intrinsic reward mechanism, i.e. humans can give themselves rewards without external stimulus rewards. In most traditional methods, such as reinforcement learning, an externally determined reward is needed. How could we design an intrinsic reward to make better intelligent systems? How can we have self-detection of error?
8. *Credit assignment.* Some researchers indicate that the learning problem is a credit assignment approach.
9. *The brain predicts sensory consequences of motor commands.* Prediction and stability. The brain predicts the sensory consequences of motor commands. The ability to predict accurately leads to stability, e.g. some mechanisms in our brain that predict the consequences of certain actions we make (e.g. Volunteer that tries to maintain holding a book. If someone else removes the book, the left hand automatically moves up. But if the volunteer himself removes the book with the right hand, the brain somehow can predict the exact timing and make auto-correction (cancellation) such that the left hand will not move up). The human brain performs predictive trajectories when playing games such as table tennis, badminton, etc.
10. *Children's brains.* After six or seven years old, the brain reach about 95 percent of the average adult volume, having a fourfold increase in size since birth. The latest research finding indicates that even so, the brain continues to change, i.e. extensive internal re-wiring continues.
11. *Neurogensis in learning and memory.* It is not yet clear how neurogenesis improves or weakens different forms of learning and memory. Hen and colleagues temper the exuberance surrounding neurogenesis research with a cautionary note. "Strategies aimed at stimulating hippocampal neurogenesis to elicit antidepressant or precognitive effects will need to strike a fine balance between restoring function and avoiding potential negative consequences of an excess of neurogenesis." — Tom Fagan.
12. *Plasticity and aging.* Part of the loss of plasticity is probably due to age. But some of the loss of plasticity might be by program. Once a neuron has been around for a while it has probably found some purpose and there's probably a bias in the brain's design against letting

a neuron too easily get reprogrammed for other purposes. (Ge *et al.*, 2007).

13. *SK gaining sight and critical period issues.* On the case of patient SK gaining sight, there could be a distinction between Hubel and Weseil's cat experiment, which was performed in total darkness compared to SK, which was brought up with exposure to light.
14. *Why does the brain need background noise to learn?* This background noise causes a gradual change in the brain's neural activities even when nothing new is being learned (Seung's lab) (Rokni *et al.*, 2007). Rokni *et al.* in their experiment show that the tuning curves of motor cortical cells are constantly changing even when performing a familiar task. Furthermore, when learning a new motor task, learning-related changes occur on top of this background of changing tuning curves. From the existence of background changes, the researchers concluded from their study of motor learning:

    a. The brain plasticity process is considerably variable.
    b. The spatial randomness of background changes: they inferred that the source of variability is local, i.e. independent in different synapses, rather than from environment noise, e.g. muscle noise, which through sensory feedback, contaminates the learning signal.
    c. Plasticity noise is additive.
    d. Noise changes synapses very slowly. According to their theory, this slowness is necessary to prevent the noise from erasing motor memories.

15. *Dopamine* also plays a critical role in many other neurological conditions, including attention deficit disorder, schizophrenia, and drug addiction. Hence, with understanding the detailed effects of dopamine in the brain, drugs could be designed to more directly target beneficial actions without producing as many unwanted side effects. The chemistry of our brain largely influences our personality and emotions. "*One personality trait in humans is how sensitive and responsive we are to incentives and rewards.*" — Richard Depue (Cornell University).
16. *Stability.* A central issue in neuroscience is how the adult brain selectively adapts to important environmental changes. Although the brain needs to adapt to new environments, its architecture

must protect itself from modification from the continual bombardment of undesirable information. How does the brain solve this so-called "stability-plasticity dilemma"? The stability and plasticity dilemma is largely unsolved. *How does the brain achieve stability and plasticity? (Stability and plasticity dilemma)*. It is widely accepted that the synaptic weights are the support of learning. In the training phase (or learning phase), the neural network system computes the weights, and the weights are then fixed during the generalization phase of the network. If synaptic weights do not stop varying, what is the nature of them when confronted to continuous sequences of inputs? Such a system will face catastrophic forgetting, which is the forgetting of old patterns due to the presentation of new ones. This is the stability and plasticity dilemma. Although there are several discussions on how to resolve the stability and plasticity dilemma, the problem persists and we do not indeed understand how.

a. How does the brain maintain stability enough for learning new information and yet retain old information or knowledge? How can a network be made flexible enough to learn and yet be stable enough to retain knowledge? Although some recent studies have shown that visual learning is susceptible to disruption and elucidates the processes by which the brain can consolidate learning and thus protect what is learned from being overwritten.

17. *Shaping behavior* is an aspect of behavior analysis that gradually teaches new behavior through the use of reinforcement until the target behavior is achieved (Wolfgang, 272).
18. *NMDA receptors* are a type of glutamate receptor, critical in long-term potentiation and found in the hippocampus, mostly CA1. It controls a calcium ion channel, which normally is blocked by a magnesium ion. Calcium is critical for long-term potentiation.
19. *AMPA* receptors control sodium channels — involved once long-term potentiation has occurred.
20. *The inferior temporal cortex* is necessary for visual pattern discrimination and receives information from the visual cortex.

21. *Extinction in operant conditioning* can occur when the behavior or response fades out over time due to non-reinforcement.
22. *Some locations are related to learning and memory in the brain.*

    a. Perrihinal cortex — Within the medial temporal lobes, the Perrihinal cortex lesions appear most significant for object recognition. This is because the Perirhinal cortex serves as the final stage in the ventral visual cortical pathway that represents stimulus features and it operates as part of a network for associating sensory inputs within and across sensory modalities.
    b. Hippocampus — The hippocampus stores a series of allocentric maps (representation of space based on external objects and landmarks) based on sensory input. The hippocampus is involved in spatial memory.
    c. The inferotemporal cortex (cortex of the inferior temporal lobe) is involved in the perception of objects and storing memories of visual patterns.
    d. Inferotemporal cortex — IT lesions cause agnosia ("psychic blindness").
    e. Prefrontal Cortex — Lesions of the lateral prefrontal cortex impair delay tasks.
    f. Striatum — It is involved with storing memories of contingently associated responses and stimuli. These are memories that are built up through multiple stimulus-response contingencies. (Fuster, 1997).
    g. Orbitofrontal cortex (OFC) — The human orbitofrontal cortex is among the least understood regions of the human brain but has been proposed to be involved in sensory integration, in representing the affective value of reinforcers, and in decision-making and expectation (Kringelbach, 2005). In particular, the human orbitofrontal cortex is thought to regulate planning behavior associated with sensitivity to reward and punishment (Bechara *et al.*, 1994). This is supported by research in humans, non-human primates, and rodents. Researchers have found that the reward value, the expected reward value, and even the subjective pleasantness of foods and other reinforcers

are represented in the OFC. A large meta-analysis of the existing neuroimaging evidence demonstrated that activity in the medial parts of the orbitofrontal cortex is related to the monitoring, learning, and memory of the reward value of reinforces, whereas activity in the lateral orbitofrontal cortex is related to the evaluation of punishers, which may lead to a change in ongoing behavior (Kringelbach and Rolls, 2004). Similarly, a posterior-anterior distinction was found with more complex or abstract reinforcers (such as monetary gain and loss) being represented more anteriorly in the orbitofrontal cortex than less complex reinforcers such as tast. It has even been proposed that the human orbitofrontal has a role in mediating subjecting hedonic experience (Kringelbach, 2005).

h. Destruction of the OFC through acquired brain injury typically leads to a pattern of disinhibited behavior. Examples include swearing excessively, hypersexuality, poor social interaction, compulsive gambling, excessive alcohol/smoking/drug use, and poor empathizing ability.

i. Decisions about the visual world can take time to form, especially when the information is unreliable. (Roitrnan and Shadlen, 2002) studied the neural correlate of gradual decision formation by recording activity from the lateral intraparietal cortex (area LIP) of rhesus monkeys during a combined motion-discrimination reaction-time task.

j. During the period of decision formation, the epoch between the onset of visual motion and the initiation of the eye movement response, LIP neurons underwent ramp-like changes in their discharge rate that predicted the monkey's decision.

23. *Hebb Learning (Hebbian Theory)* Donald Hebb extended what Cajal proposed on synaptic strength connection to include that cells may grow new connections among each other to communicate and form new memory. Hebb described a mechanism for synaptic plasticity. This mechanism was known as Hebbian theory or also called Hebb's rule, which indicates that an increase in synaptic efficacy arises from the pre-synaptic cell's repeated and persistent stimulation of the post-synaptic cell. The theory is often summarized as "cells that fire together, wire together".

Chapter 6

1. The importance of feeling understood is emphasized by Stephen Covey in his bestselling book, "The 7 Habits of Highly effective people" (Covey, 1989).
2. Multiple emotion systems. There are still issues with regard to how the emotion memory (after being captured) is exhibited in other brain regions. The current observation suggests that the brain may have multiple emotion systems, e.g. multiple fear systems. Professor David Amaral's research on amygdala lesions in the neonatal monkey produced decreased fear to inanimate objection but increased fear in novel social interaction. The issue is why is there an increased fear in novel social interaction? Could it be due to a multiple fear system in the brain? He also observed that amygdala lesions affect emotion (specifically fear) learning and expression. However, in monkeys, it was shown that it affects fear learning but not the fear "startle". For example, in a monkey and rat on "fear potential startle" training such as turning on a light and getting an electric shock. The monkey and rat will learn that when the light is on, they will get an electric shock. And hence, when the light is on, the monkey or rat will jump even if there is no electric shock. For the rat, if the amygdala is removed, the condition learning is also removed and there is no fear "startle". But in the monkey, the condition still exists and that leads to the question of where the condition is stored in the brain.
3. "*There is no **fear** in **love**. But perfect **love** drives out **fear**, because **fear** has to do with punishment. The one who **fears** is not made perfect in **love**.*" — 1 John 4:18.
4. Losses of the ventromedial prefrontal regions (located on the lower forward section of the frontal lobe). Studies of those who have sustained damage to this region indicate that they seem to exhibit the same sort of behavior change. They seem unable to understand the need for long-range planning, and thus to act irresponsibly.
5. In (Tropea *et al.*, 2006), the experiment shows that activity-dependent development of the primary visual cortex (V1) involved in DR (dark-rearing, i.e. reduction of activity in both eyes via dark-rearing) leads to the upregulation of genes subserving synaptic transmission and electrical activity.

6. A simple rule of how the OCC model will synthesize the emotion "joy" is:

   If $E > 0$
   Then set $P_j = f_j(E, I)$

   E is greater than 0 if the event is expected to have beneficial consequences. And I represents a combination of global intensity variables (e.g. expectedness, reality, proximity). Then P_j is the potential for generating a state of "joy" where f is functioning specific to "joy". Similar rules can be used for computing potentials for other emotions, e.g. the potential for distress, P_d.

   This rule activates the "joy" emotion to give rise to a value (P), which can be mapped to one of a variety of emotion terms in the "joy" family, such as "pleased" for a moderate value or "euphoric" for an unusually high value. Note that the example of the emotion joy is an oversimplified case; more complicated rules exist for other emotional types in the OCC model.
7. *Cathexis — four elicitors for emotion synthesis.* The updating rule contains terms that take on values specific to proto-specialists, but otherwise the form is the same for every proto-specialist's emotion intensity. At each time t, each proto-specialist $p = 1 \ldots p$ updates its emotional intensity $I_p(t)$ as follows:

   - Let $E_{p,i}$, $I = 1, 2, 3, 4$ be values contributed to proto-specialist p by the four elicitors.
   - Let $\alpha_{p,m}$ be the excitatory gain applied by proto-specialist m to proto-specialist p.
   - Let $\beta_{p,m}$ be the inhibitory gain applied by proto-specialist m to proto-specialist p.

   Finally, let f be a function that controls the temporal decay of an emotion intensity, and let g be a function that constrains the emotion intensity to lie between zero and its saturation value. The new intensity is then a function of its decayed previous value, its elicitors, and influences from other emotion intensities:

$$I_p(t) = g(f\ (Ip(t-1)) + \sum_{l=1}^{4} \varepsilon_{p,l} + \sum_{m=1}^{p} (\alpha_{p,m} - \beta_{p,m}) I_m(t))$$

As in the OCC model, the intensity is compared to an emotion-specific activation threshold before determining if an emotion exists. Only if the intensity exceeds the activation threshold does the proto-specialist release its value to influence the behavior system and other proto-specialists.

8. Dolores Canamero, at the Free University of Brussels, built a system in which emotions trigger changes in synthetic hormones, and in which emotions can arise as a result of simulated physiological changes (Canamero, 1997).
9. Low-level signals are motions around the mouth and eyes, hand gestures, pitch changes in the voice, and verbal cues such as the words being used.
10. Signals are any detectable changes that carry information or messages. Sounds, gestures, and facial expressions are signals that are observable by natural human senses, while blood pressure, hormone levels, and neurotransmitter levels require special sensing equipment.
11. High-level reasoning and low-level signals cooperate in the generation of emotional expression.

Chapter 7

1. Solving the noise saturation dilemma. Suppose that the ith cell C_i receives an input I_i that can turn on some of its B excitable sites by mass action. Let $x_i(t)$ be the number of excited sites and $B-x_i(t)$ be the number of unexcited sites at time t. The simplest mass action law for turning on unexcited sites and letting excited sites spontaneously turn off is:

$$\frac{d}{dt}x_i = -Ax_i + (B - x_i)I_i,$$

where $I = 1, 2, \ldots, n$. Term $(B-x_i)\, I_i$ says that the input I_i turns on unexcited sites $B-x_i$ by mass action. Term $-Ax_i$ says that excited sites spontaneously become unexcited by mass action at rate A. Hence, when $I_i = 0$, x_i can decay to the equilibrium point zero.

2. *Recurrent on-center off-surround networks.* The system equation:

$$\dot{x}_i = -Ax_i + (B - x_i)f(x_i) - x_i \sum_{k=i} f(x_k) + I_i,$$

where $i = 1, 2, \ldots, n$, and $x_i (\leq B)$ is the mean activity of the *i*th cell or cell population, v_i, of the network. Four effects determine this system: (1) exponential decay via the term $-Ax_i$; (2) shunting self-excitation via the term $(B-x_i)f(x_i)$; (3) shunting inhibition of other populations via the term $-x_i\Sigma f(x_k)$; (4) externally applied inputs, via the term I_i. The function $f(w)$ describes the mean output signal of a given population as a function of its activity w. *In vivo* $f(w)$ is often a sigmoid function of w (Grossberg, 1973). The recurrent on-center off-surround networks is said to take into account the non-linearity, shunting, on-center off-surround feedback interaction.

3. *Hodgkin-Huxley's cell membrane equation and Grossberg's shunting equation.* The shunting equation is claimed to be compatible with the (Hodgkin and Huxley, 1952) cell membrane equation. Below is a comparison of the two equations.

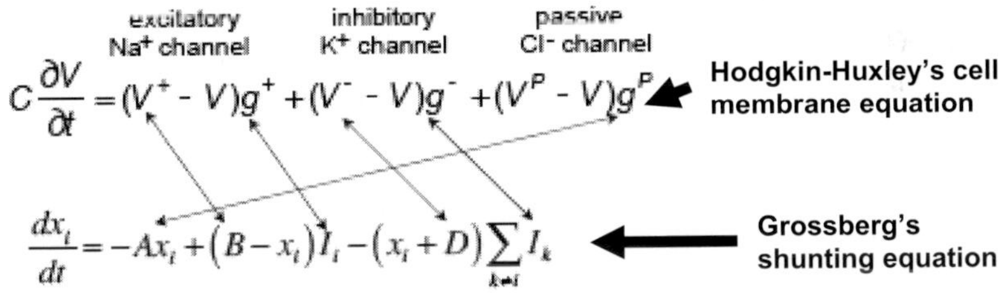

where
$$V^+ \to B$$
$$C \to 1$$
$$V^- \to D$$
$$V^P \to 0$$
$$g^P \to A$$
$$g^+ \to I_i$$
$$g^- \to \sum_{k=i} I_k$$

For the shunting equations, $-Ax_i$ is the passive decay term, I_i is the excitatory input, and I_k is the inhibitory input. Cell activity is bound between D and B, independent of the excitatory input (I_i) and inhibitory input (I_k).

4. Equation on the learning law. Grossberg in 1988 discussed the learning law for an adaptive weight or long-term memory trace $z(t)$ (also sometimes denoted by $w(t)$ or $m(t)$):

$$\frac{dz}{dt} = f(x)[-Az + g(y)],$$

where x is the activity of a presynaptic (or postsynaptic) cell, y is the activity of a postsynaptic (or presynaptic) cell, and z is the adaptive weight at the synapse of an intervening pathway or axon. This apparently simple equation was, for example, used to introduce self-organizing maps, adaptive resonance theory and counter-propagation networks. This simple equation and its variant gives rise to such varied names as the outstar learning law, the instar learning law, the gated steepest descent law, Grossberg learning, Kohonen learning, and mixed Hebbian/anti-Hebbian learning. Read the (Grossberg, 1998) paper for more discussion.

5. *ART model.* ART stands for Adaptive Resonance Theory. It was first introduced in 1976 by Grossberg as a theory of human cognitive information processing (Grossberg, 1976), and later, the theory evolved as a series of real-time neural network models that perform unsupervised and supervised learning, pattern recognition and prediction (Carpenter and Grossberg, 1987) (Carpenter and Grossberg, 2003). The basic ART system is an unsupervised learning model. Many other models evolved from ART, two of which are fuzzy ART for analog input patterns (Carpenter *et al.*, 1991) — ARTMAP models for combining two unsupervised modules to carry out supervised learning (Carpenter *et al.*, 1992). Many variations of the basic ART supervised and unsupervised networks have since been adapted for technological applications and biological analyses. Some of these are further discussed below.

ART proposes how the brain learns to recognize objects and events. This is accomplished through an interaction between bottom-up perceptually-driven inputs and learned top-down expectations. Bottom-up inputs attempt to match top-down expectations and the top-down expectations can prompt the brain to anticipate input or expectation patterns. When a match occurs, the system locks into an attentive resonant state that drives the recognition learning process (how we learn to recognize things), and hence, the term adaptive resonance. ART predicts that all conscious events are resonant events. The degree of match that is required for resonance and sustained attention to occur is set by a vigilance parameter. Vigilance may be increased by predictive errors, and controls whether a particular learned representation will be concrete or abstract. Low vigilance allows the learning of broad abstract recognition categories; high vigilance forces the learning of specific concrete categories. If a match is inadequate, then the current input is processed as a novel stimulus. Attention is then rapidly reset so that memory can be searched for another, or new, representation of the event. (Note: The iSTART model proposes that individuals with autism have their vigilance fixed at such a high setting that their learned representations are very concrete or hyperspecific; that is, hypervigilance leads to hyperspecific learning. This property leads to a multitude of problems with learning, cognition, and attention due to the manner in which top-down matching, attention, learning, attention focusing and reset, and memory search are organized. Grossberg indicated that the ART model clarifies thalamo-cortical-hippocampal interactions among others in the brain.) ART is born of many other models, to follow in the quest of CNS's group in modeling the brain.

Grossberg addressed how ART solved the stability-plasticity dilemma using dynamic balance. He demonstrated using ART, and explained how the brain dynamically switches between its stable and plastic modes such that it is not too rigid (due to being too stable) or chaotic (due to being too plastic).

Grossberg also explained that the ART resonant state synchronises, amplifies, and prolongs the cell activations that it includes. This change of energy and time scales enables the more slowly varying adaptive weights to learn the activities that they can correlate. By discarding signals other than those that are attended, the resonant state helps to prevent catastrophic forgetting. It is because the resonance triggers learning that the theory is called the adaptive resonance theory (Grossberg, 2000b).

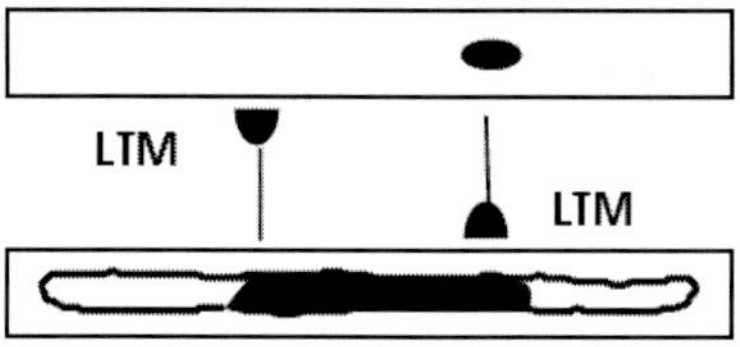

The above figure shows an example of recognition categories when a subset of cells is selected by active top-down prototypes; they can continue to send bottom-up signals to the category representations at the next processing stage. The categorical cells can, in turn, continue to send top-down signals to the attended feature pattern. A resonant state hereby locks in, whose cells are synchronized, amplified, and prolonged long enough for the LTM traces that support the resonant cells to learn from their activity patterns.

ART matching and resonance rules have the bottom-up activation (by itself ART can automatically activate target nodes) and top-down expectations. The top-down expectations can learn prototypes that select consistent bottom-up signals, suppress inconsistent bottom-up signals (attentional focusing), and cannot by themselves fully activate target nodes (modulation, priming).

Carpenter indicated that ART principles have further helped explain parametric behavioral and brain data in the areas of visual perception, object recognition, auditory source identification, variable-rate speech, and word recognition (Carpenter and Grossberg, 2003). More recent progress has also been made on how the neocortex is organized into layers, clarifying how ART design principles are found in neocortical circuits.

6. *CogEM model* stands for the Cognitive-Emotional Motor. The CogEM model proposes how emotional centers of the brain, such as the amygdale, interact with sensory and prefrontal cortices (notably ventral or orbital, prefrontal cortex) to generate affective states. Grossberg indicated that CogEM explains the interaction between cognitive, emotional, and motor learning properties (Grossberg, 2000). The CogEM model was first introduced in (Grossberg, 1971) and has since undergone substantial development in later years. See (Grossberg, 1982a, 1982b, 1984b), (Grossberg and Gutowski, 1987), (Grossberg and Levine, 1987), (Grossberg and Merrill, 1992, 1996). CogEM extends ART to the learning of cognitive-emotional associations or

associations that link objects and events in the world to feelings and emotions that give these objects and events value. Under normal circumstances, arousal of the circuits in the brain that control emotion are set at an intermediate level. Either under-arousal or over arousal of these circuits can cause abnormal emotional reactions and problems with cognitive emotional learning.

> "*If the emotional center is over-aroused, the threshold to activate a reaction is abnormally low, but the intensity of the emotion is abnormally small ... In contrast, if the emotional circuits are under-aroused, the threshold for activating an emotion is abnormally high, but when this threshold is exceeded, the emotional response can be over reactive.*" said Grossberg. The iSTART model proposes that individuals with autism experience under aroused emotional depression, which helps explain symptoms like reduced emotional expression as well as emotional outbursts.

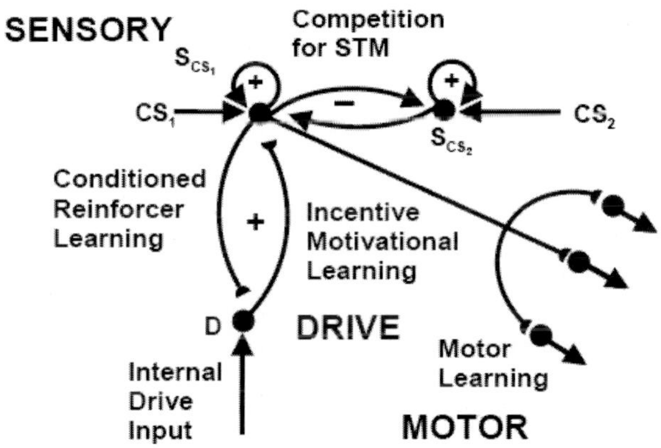

The figure shows the simplest CogEM model. Three types of interacting representations (sensory, drive, and motor) that control three types of learning (conditioned reinforcer, incentive motivational, and motor) help to explain many reinforcement learning data. Sensory representations, "S", temporarily store internal representations of sensory events in working memory. Drive representations, "D", are sites where reinforcing and homeostatic, or drive, cues converge to activate emotional responses. Motor representations, "M", control the read-out of actions. Conditioned reinforcer learning enables sensory events to activate emotional reactions at drive representations. Incentive motivational learning enables emotions to generate a motivational set that biases the system to process information consistent with that emotion. Motor learning allows sensory and cognitive representations to generate actions (Grossberg, 2000) (Grossberg and Seidman, 2006).

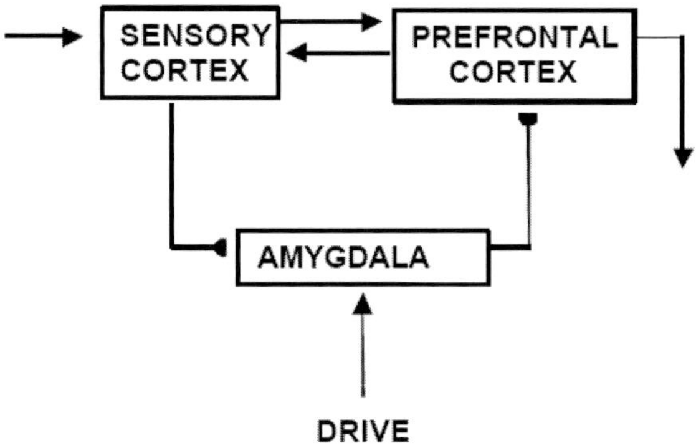

The figure shows a block diagram of a CogEM model in terms of simplified anatomical connections. When a drive representation like the amygdale gets depressed, its diminished activation by sensory events prevents normal interpretation of emotionally important events and also attenuates motivationally appropriate signals to and from the prefrontal cortex. Such a local imbalance in the model circuit can generate many negative symptoms that are characteristic of schizophrenia, including the loss of a theory of mind.

7. START model. A synthesis of the ART and CogEM models, called the START model, stands for Spectrally Timed ART Model (Fiala *et al.*, 1996) (Grossberg and Merrill, 1992). The START model predicts how motivational mechanisms within the amygdala and related emotion-representing brain areas can rapidly draw motivated attention to salient cues. This can happen via a cognitive-emotional resonance within CogEM feedback circuits between sensory representations "S" and drive representations "D".

   The degree of match that is required for resonance to occur is set by a vigilance parameter, which controls whether a particular learned representation will be concrete or abstract. Low vigilance allows for the learning of broad abstract recognition categories, such as a category that is activated by any face; high vigilance forces the learning of specific concrete categories, such as a category that is activated by a particular view of a familiar friend's face.

8. *iStart model*, which stands for Imbalanced Spectrally Timed Adaptive Resonance Theory, is derived from the earlier START model developed by Grossberg to explain how the brain controls normal behaviors. The new model illustrates how brain mechanisms that control normal emotional, timing, and motor processes may become imbalanced and lead to symptoms of autism. START and its imbalanced version, iSTART, are a combination of three models, each of which tries to explain fundamental issues about human learning and behavior (Grossberg and Seidman, 2006).
9. *ARTSCAN* demonstrates the theoretical analysis of how spatial and object attention interact with surface and boundary representations to support the learning of view-invariant object categories (Fazl *et al.*, 2005) (Fazl *et al.*, 2006). The ARTSCAN model predicts how spatial and object attention work together as eye movements scan a scene to selectively fuse, through learning, multiple views of an object into a view-invariant object category.
10. *ARTSTREAM*. A neural network model of auditory scene analysis. This model proposes how the brain accomplishes this feat. The model clarifies how the frequency components that correspond to a given acoustic source may be coherently grouped together into a distinct stream based on pitch and spatial location cues. The model also clarifies how multiple streams may be distinguished and separated by the brain (Grossberg *et al.*).
11. *LIGHTSHAFT* stands for the LIGHTness and SHApe-From-Texture model. This is a model that utilizes monocular visual texture information to produce a 3D percept of surface lightness. It is built upon and extends the FAÇADE model (Grossberg *et al.*, 2007).
12. *The ARTEX* model is proposed by (Grossberg and Williamson, 1999). The model joins together visual preprocessing (multiple-scale bottom-up filtering, horizontal grouping, and surface filling-in) as a perceptual front end to an ART classifier to learn and categorize both Brodatz textures (based on (Brodatz, 1966) benchmark work on

texture) and natural textured scenes after they were processed by a synthetic aperture radar (SAR) sensor.

Chapter 8

1. **Bayesian Network Knowledge Fragments Fusion**

This section discusses the technical details of how to fuse BN knowledge fragments. We present two novel methods to handle the fusion of multiple Bayesian Network knowledge fragments, which we termed N-Combinator and N-Clone.

The problem of BN knowledge fragments fusion is depicted in the example given in Fig. A1. In this example we wish to fuse fragments 1 and 2 encoded with the parameters $P(A|B)$ and $P(A|C)$ respectively to form the combined fragment where node A now has multiple parents, namely, B and C. This fusion process is needed in D'Brain as it allows the Situation Specific Bayesian Network to be created from the library of knowledge fragments.

One approach is to compute the higher dimension CPT of $P(A|B,C)$, from lower dimension CPTs $P(A|B)$ and $P(A|C)$. There exists a widely known method of CPT combination called Noisy-OR if all the nodes in the model are binary (Pearl, 1988). Noisy-OR is usually used to describe the interaction between n causes x_1, $x_2,\ldots, x_n$ (parent nodes) and their common effect y (child node). The causes x_i are each assumed to be sufficient to cause y in absence of other causes and their ability to cause y is assumed to be independent of the presence of other causes. Here, we do not consider leak probabilities where the model does not capture all possible causes of effect node y. The leak probability models the situation

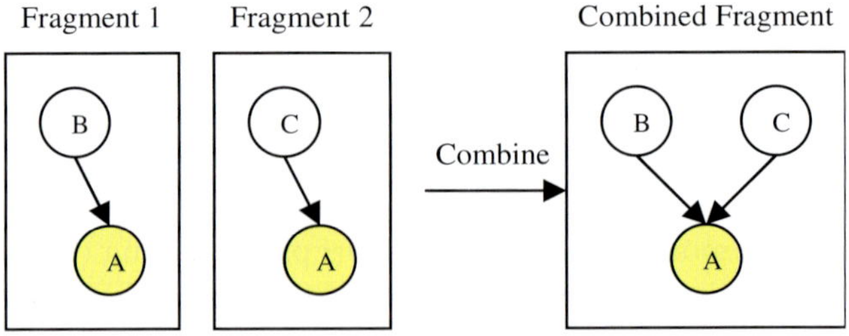

Fig. A1. Graphical Representation of Problem Statement.

where the effect node y will be active even when all its causes are inactive.

The mathematical formulation for the probability that the effect node y is active, given a subset of its active parents x_p is:

$$P(y \mid x_p) = 1 - \prod_{i:x_i \in x_p} (1 - p_i) \tag{1}$$

where p_i is defined as:

$$\begin{aligned} p_i &= P(y \mid \overline{x_1}, \overline{x_2}, \ldots, \overline{x_{i-1}}, x_i, \overline{x_{i+1}}, \ldots, \overline{x_{n-1}}, \overline{x_n}) \\ &= P(y \mid x_i) \end{aligned} \tag{2}$$

which is the probability that the effect node y is active given a single parent x_i is active. Note that $\overline{x_i}$ denotes that x_i is not active.

While Noisy-OR is a widely accepted CPT combination method, it is limited to only binary nodes modeling. Diez extended the Noisy-OR to a Noisy-Max model that allows the computation of full CPTs for multiple state nodes but on the condition that the nodes (effect node and all its parents) have to be graded. This means that if a node has n states, its nth state has to be greater in its effect than its $(n - 1)$th state, which is greater in its effect than its $(n - 2)$th state and so on. For example, if the node is Illness, its graded states can be None, Mild, Serious, and Critical. On the other hand, if the node is Color and its states are Red, Green, and Blue, then the node Color is not graded since we cannot say that the effect of Red > Green > Blue.

Noisy-OR is not a suitable method for fusing BN knowledge fragments due to the restriction that only binary nodes are used in the knowledge fragments. Modeling complex situation using only binary nodes is inadequate. Although Noisy-Max allows us to model nodes with more than two states, it requires nodes to be graded. However, it is not always possible to model graded nodes for some applications. Hence, Noisy-Max is also not a suitable method. The absence of an appropriate method for fusing BN knowledge fragments for generic nodes (multi-state nodes with no restriction on their states) provided the motivation for us to propose two novel methods: N-Combinator and N-Clone.

2. **N-Combinator**

The N-Combinator computes the joint CPT entries for a child node with multiple parent nodes without the constraint that they have to be graded. There are three steps to the N-Combinator method.

Step 1: Model each multi-state child node and its multi-state parent nodes as multiple binary nodes.
Step 2: Combine the binary nodes using Noisy-OR.
Step 3: Re-model the binary nodes back to their original multi-state nodes by normalization.

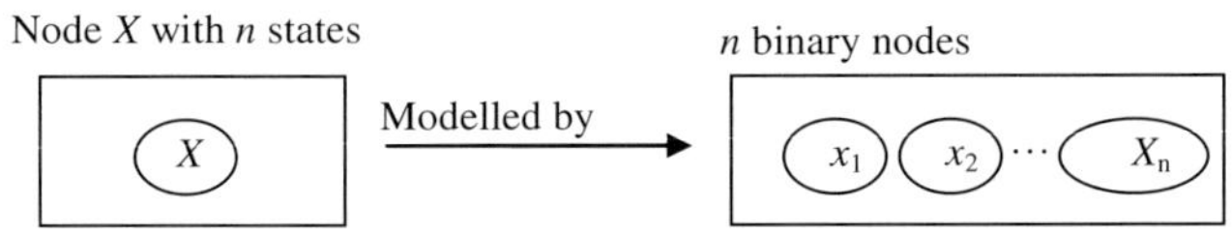

Fig. A2. Modeling a multi-state node with multiple binary nodes.

Consider the node X in Fig. A2 with n states $\{x_1, \ldots, x_n\}$. In Step 1, we model it into n binary nodes where each binary node has an active and non-active state represented by x_i and $\bar{x}_i$ respectively. Then we have

$$(X_i = x_i) \equiv (\overline{x_1}, \ldots, \overline{x_{i-1}}, x_i, \overline{x_{i+1}}, \ldots, \overline{x_n}) \quad (3)$$

To illustrate how Step 2 and Step 3 are carried out, we use the example of computing the joint CPT entries for a child node with two parent nodes shown in Fig. A3. In this example, we model the multi-state nodes *A*, *B*, and *C* (each multi-state node has three states) into the binary nodes a_1, a_2, a_3, b_1, b_2, b_3, c_1, c_2, c_3 (Step 1). Next, these multiple binary nodes are combined via Noisy-OR (Step 2) using $P(A|B)$ (stored in fragment 1) and $P(A|C)$ (stored in fragment 2), and the Noisy-OR conditional probabilities are normalized

(Step 3) to yield the full CPT of $P(A|B,C)$. For example, (refer to Fig. A4 for illustration),

$$\begin{aligned} &P(A = a_3 \mid B = b_1, C = c_2) \\ &\quad \equiv \alpha\ P(a_3 \mid b_1, \overline{b_2}, \overline{b_3}, \overline{c_1}, c_2, \overline{c_3}) \\ &\quad = \alpha\ [1 - (1 - P(a_3 \mid b_1)) \cdot (1 - P(a_3 \mid c_2))] \qquad (4) \\ &\qquad\qquad \text{(Using Eqs. (1) and (2))} \end{aligned}$$

where $\alpha = \dfrac{1}{\sum_{i=1}^{3} P(a_i \mid b_1, \overline{b_2}, \overline{b_3}, \overline{c_1}, c_2, \overline{c_3})}$ is the normalizing factor.

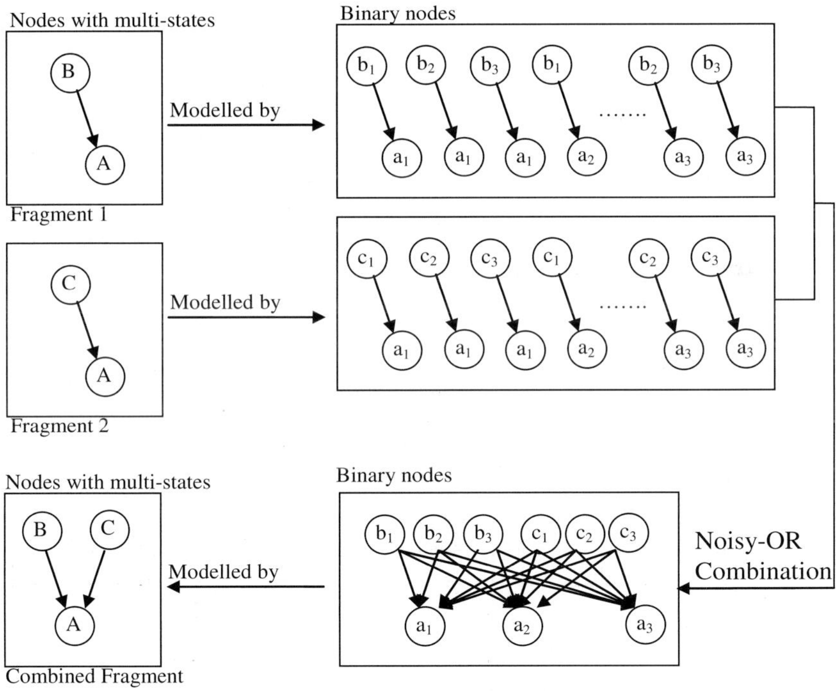

Fig. A3. Example of N-Combinator method.

The general equation for computing the joint CPT of a child node Y that has n parent nodes $X_1, X_2, \ldots, X_n$ (Fig. A5) is

$$P(Y = y_j \mid X_1 = x_1, X_2 = x_2, \ldots, X_n = x_n) = \alpha \left[1 - \prod_{i=1}^{n} (1 - P(y_j \mid x_i)) \right] \quad (5)$$

where $\alpha = \dfrac{1}{\sum_{j=1}^{m} P(Y = y_j \mid X_1 = x_1, X_2 = x_2, \ldots, X_n = x_n)}$, and

m is the number of states of Y.

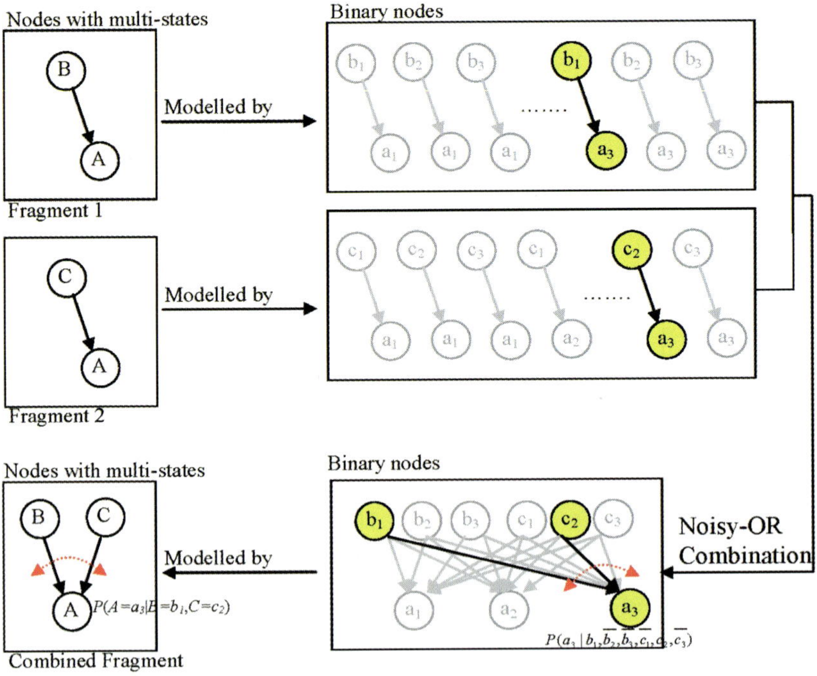

Fig. A4. Illustration of obtaining $P(A = a_3 | B = b_1, C = c_2)$.

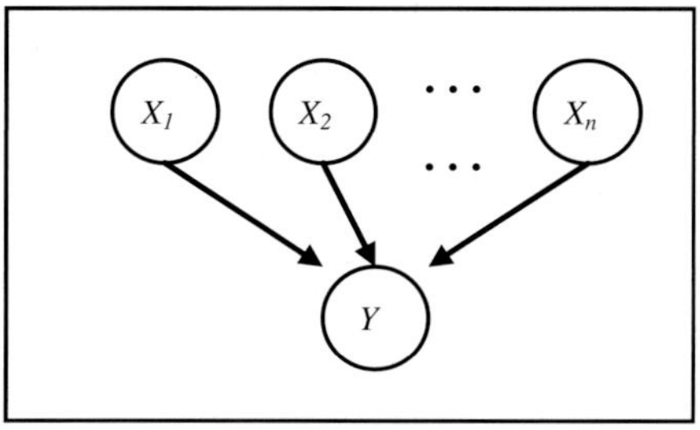

Fig. A5. Child node Y, with n parent nodes $X_1, X_2, \ldots, X_n$.

3. N-Clone

Recall that a BN is fully specified by its network structure and the corresponding network parameters (CPTs). In the N-Combinator approach to BN fragment fusion, we effectively fused the network structures of all fragments into one, which requires the higher dimension joint CPT to be computed. On the other hand, if we can somehow maintain the network structure of the individual fragments in the fused network, then there is no need to compute new CPT parameters. This is the approach adopted in N-Clone. In N-Clone, the child node is cloned. In this way, we do not need to generate the joint CPT entries. In doing so, we avoid the problem of poor inference results from poorly generated joint CPT entries.

Let's suppose we have these two fragments:

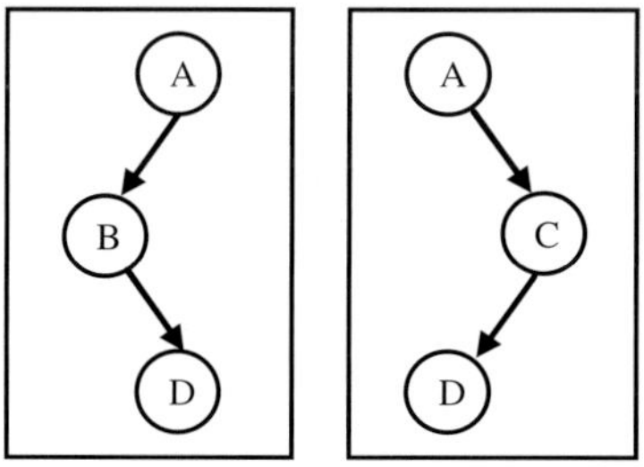

Instead of combining node D's parents in N-Combinator to form

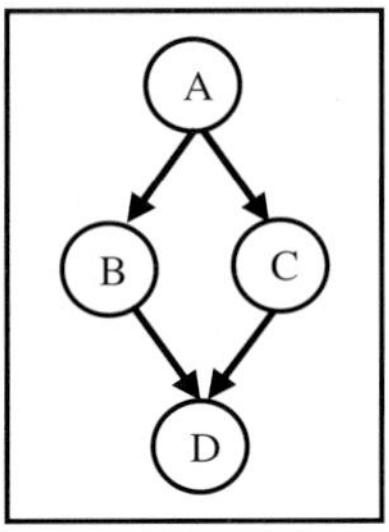

we now clone the node D:

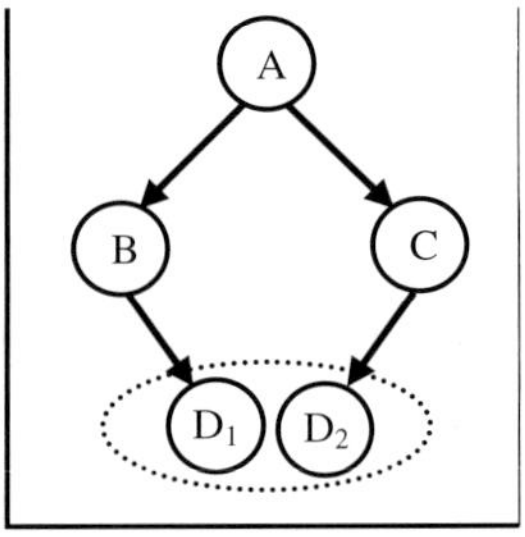

where D_1 and D_2 are the clone nodes. They are kept as separate nodes in D'Brain. However, in the graphical user interface, the user will not view the clone nodes. To the user, all the clone nodes will be displayed as one single node. This is as though the two nodes were merged into a single node as shown by the dotted ellipse. Again, note that we do not need to generate joint CPTs with this method:

$P(D_1|B)$ is obtained from $P(D|B)$ in fragment 1;
$P(D_2|C)$ is obtained from $P(D|C)$ in fragment 2.

Fragment 1 gives us some information about node D (now cloned as D_1). Fragment 2 also gives us some information about node D (now cloned as D_2). The posterior of node D is computed by combining the posteriors of node D_1 and D_2. For example, if A is in state 1, we have a posterior probability for node D_1, and a posterior probability for node D_2. Let's suppose their posterior probabilities are as follows:

Prob(D_1 = State 1)	0.3
Prob(D_1 = State 2)	0.7

Prob(D_2 = State 1)	0.4
Prob(D_2 = State 2)	0.6

We propose to combine these two posterior probabilities by averaging:

Prob(D = State 1)	(0.3 + 0.4)/2 = 0.35
Prob(D = State 2)	(0.7 + 0.6)/2 = 0.65

This is the posterior probability that the user will view if he queries about node D (i.e. given evidence that node A is in state 1, the probability that node D will be in state 1 is 0.35).

Chapter 9

1. Wordnet: http://en.wikipedia.org/wiki/Wordnet
2. Cyc: http://en.wikipedia.org/wiki/Cyc
3. Open Mind Common Sense: http://en.wikipedia.org/wiki/Open_Mind_Common_Sense

    a. The Open Mind project's goal is to teach computers all those things an average person knows but takes for granted, because they are so obvious. This is known as the problem of giving computers "common sense". This includes things like:

        (1) Every person is younger than their mother.
        (2) You can push something with a straight stick.
        (3) One hundred dollars is a lot to pay for a sandwich.
        (4) Snow is cold and is made of millions of snowflakes.
        (5) A week is longer than a minute.
        (6) Computers need a source of power to operate.
        (7) Most birds can fly, except for penguins and birds with broken wings.

4. *Mindpixel*, started in 2000, was conceived by the late Chris McKinstry, a computer scientist. Mindpixel aimed to create a database of millions of human validated true/false statements or probabilistic propositions. Participants in the project created one-line statements that aimed to be objectively true or false to 20 other anonymous participants. In order to submit their statement they had first to check the true/false validity of 20 such statements submitted by others. Participants whose replies were consistently out of step with the majority had their status downgraded and were eventually excluded. Likewise, participants who made contributions that others could not agree were objectively true or false had their status downgraded. A validated true/false statement is called a mindpixel. The project enlisted the efforts of thousands of participants and claimed to be "the planet's largest artificial intelligence effort".

    a. McKinstry believed that the Mindpixel database could be used in conjunction with a neural net to produce a body of human "common sense" knowledge that would have *market value*. Participants in the project are awarded shares in any future value according to the number of mindpixels they have successfully created.
    b. On 20 September 2005, Mindpixel lost its free server and is no longer operational. It was being rewritten by Chris McKinstry as Mindpixel 2 and was intended to appear on a new server in France. Chris McKinstry committed suicide on 23 January 2006 and the future of the project and the integrity of the data is uncertain.
    c. Some Mindpixel data (the 80K set) was utilized (in 2006) by Michael Spivey of Cornell University and Rick Dale of The University of Memphis to study theories in high-level reasoning and continuous temporal dynamics of thought.

5. *GPS (General Problem Solver)*. One of the very earliest AI programs, Newell, Shaw, & Simon's (1960a) General Problem Solver (or GPS) could be seen as a collection of critics. The critics of GPS were essentially reactive, in that they recognized differences between the system's goals and the present situation, and immediately took actions to reduce those differences; John McCarthy referred to the GPS as a

"symbolic servo-mechanism." GPS did not anticipate the consequences of those actions, or otherwise consider their relative merits; it did not engage in the kinds of hypothesis-manipulation forms of deliberation discussed in this thesis. However, in (Newell, Shaw, & Simon, 1960b) they describe what could be the first AI system with a reflective level. The system consisted of two separate GPS systems, one that operated on the base-level problem domain, and the second that operated on the knowledge base of the first GPS. This "second-order GPS" consisted of a set of critics that modified the critics of the first-order GPS so that their effects were more orthogonal and interfered with each other less. This is a form of reflection that EM-ONE does not presently engage in — reformulating the contents of its knowledge base in order to improve and accelerate inference. In EM-ONE, such an operation would probably be part of the self-reflective level, as it operates not so much on traces of recent deliberation but rather on the entire knowledge base of critics used by the system.

6. *EM-ONE.* The EM-ONE knowledge representation scheme was originally inspired by the Cyc upper level ontology. EM-ONE aspires to eventually support the use of large bodies of commonsense knowledge, and I considered using the Cyc ontology in EM-ONE. It was the only representation I had encountered that provided a sufficiently broad vocabulary to take the kinds of narratives and critics that I had been expressing pre-formally in English and in a sharper representation that a program could more easily work with. In the end, I decided to use my own, simpler scheme, although in the future I may try to implement a version of EM-ONE on top of the Cyc substrate. This should not be difficult because EM-ONE is built on top of a Lisp-based version of Prolog that makes use of Prolog's database and pattern matching machinery. Cyc provides this same functionality, but in many ways is more sophisticated because it has special purpose inference procedures for certain special types of commonsense inference (e.g. taxonomic and temporal inference are implemented with special-purpose algorithms).
7. Some of the differences between EM-ONE and Cyc are:

   a. In EM-ONE, there are a limited collection of frames and frame slots that let one describe physical, social, and mental situations.

This simple representation scheme is far easier to learn and apply than the full Cyc ontology, and the cost in precision and expressiveness trades off against Cyc's complexity. One of my goals at the outset of EM-ONE was to build a system that was reasonably easy for someone new to the project to understand, and I wanted to avoid putting them in the position of having to first learn the full Cyc ontology. In retrospect, the Cyc ontology is so well documented that it may have been a better decision to have selected an appropriate subset of Cyc. On the other hand, because the EM-ONE representation is completely frame-based, and because every frame is a reified entity that can be attached to slots of other frames, it is especially natural to represent mental notions such as propositional attitudes (believes, desires, etc) that relate actors to mental states, which is important for a system that does social reasoning and that can reflect upon itself and its own reasoning.

b. The EM-ONE narrative corpus is based not on default rules but instead on narratives annotated by the causal and dependency relationships that exist between the elements of the narrative. These narratives are structured in a fairly uniform manner where each narrative is a set of situations whose constituents have simple causal and temporal interrelations. Singh believed that narratives are generally a better way to represent commonsense knowledge than abstract logical rules, as is the convention in Cyc. Rather than reasoning by making deductions using commonsense rules, EM-ONE reasons by making analogies to commonsense narratives. While Cyc does have some knowledge in the form of stories and scripts, the bulk of the knowledge in Cyc is the form of general facts and rules.
c. Cyc could be thought of as having a somewhat uniform deliberative layer, but like most present day inference systems it does not possess a reflective layer. Cyc possesses some knowledge about folk psychology and mental states, but it does not employ any of this knowledge to assess and debug its own functioning. Novel non-folk-psychological concepts about the mind, such as mental critics, do not appear. This seems to be because Cyc consists primarily of types of knowledge that AI practitioners are largely already familiar with, but theories about human commonsense

psychology — ones that let us make detailed predictions about how people learn, reason, and reflect — are still in their infancy, so there was little existing research for the developers of Cyc to draw upon.

8. *Story-based intelligence* was proposed by Roger Schank. Here is a short description of his writing: Why do people like to tell and hear stories? Story telling is common enough phenomenon. But do people who like to tell stories really spend any time at all listening to other people's stories? When people who like to tell stories are listening to the stories of others, you often see them chafing at the bit. They are waiting for the story to be over so they can tell their own story. If you watch enough of this, hear enough of it, and participate enough in it by examining your own behavior, you begin to wonder if anybody is listening to anybody. The only time you really have to listen is when you attend a lecture and you cannot get up and tell your story back. Even then, your mind is going all the time, getting reminded of your own experiences that you would just love to tell if it were socially permissible to do so.
9. Knowledge Machine (KM) is a knowledge representation and reasoning system. KM is developed by the Knowledge Systems Research Group of the University of Texas at Austin, led by Peter Clark and Bruce Porter. It claims to be a powerful, frame-based language with clear first-order logic semantics. And it contains sophisticated machinery for reasoning, including selection by description, unification, classification and reasoning about actions using a situations mechanism. It is implemented in Lisp and runs on both Windows and Unix operating systems. Main website of KM: http://www.cs.utexas.edu/users/mfkb/RKF/km.html

   a. When someone tells you a problem, most of the time they do not want an answer, but a good story to relate their problem.
   b. On AI — Schank said: "The kind of intelligence you really worry about, after all, when you are sitting next to the guy in the airplane, is whether he can hold up his end of the conversation in such a way so as not to bore you to death. You see him as being really intelligent if the stories he tells back to your stories actually

relate to your stories; that is, if it seems like he's understood your stories" … "Intelligence revolves a lot around getting a machine to have something to say and getting it to have the ability to understand what you said. Intelligence involves a lot less problem solving than we commonly think."

c. "On the one hand, it is really a lot simpler to state one's belief than to tell a story that explains why one holds that belief. On the other hand, it is never as persuasive to do so. You can say, I believe such and such, and no one listens to you. If you tell a really good story, however, people listen."

d. Schank added that by telling our stories, we get reminded and we try to learn. As we tell a story to someone, we hear a story back in our minds that we like to tell. We start to tell the story because we are actually in some sense comparing the story that comes to mind to the story that was told to us. This could be called the understanding of the understanding cycle. In the understanding cycle, we start with expectation failures. We are bundles of expectations. You expect all kinds of things to happen: you expect me to stand up here and lecture. You do not expect me to dance on stage and take my clothes off. As you begin to explain the behavior of the things around you, you begin to formulate expectations: You expect the world to exist in a certain way, and if it doesn't exist that way, you worry about it. You have an expectation, and when it fails you may be upset. You may think that failure is a bad thing, but it turn out that failure is a wonderful thing. Failure is the beginning of learning. When you feel you want to get something but don't get it you get upset. What your mind is actually saying, however, is, "Oh boy! I failed. That is terrific, because now I can begin to figure out why." … The reason to call that memory a story is that it tends to have story-like properties. Story-Based Memory by Roger Schank (Schank, 1992).

Chapter 10

1. *Conductance-based models* (Skinner, 2006). The conductance-based model is a form of the Hodgkin-Huxley cell membrane model. A conductance-based model represents a neuron by a single isopotential

electrical compartment, neglects ion movements between subcellular compartments, and represents only ion movements between the inside and outside of the cell. To form a very small region of the brain, one would need to build many multiple compartments using the conductance-based model.

2. *CCortex-based autonomous cognitive model* (ACM). ACM is built by a company called Artificial Development (California-based company), a company that started modeling the entire brain around 2002/3. The company claims it has completed a representation of a functioning human brain. And it hopes that their software may have immediate applications for data mining, network security, search engine technologies, and natural language processing. ACM claims to use realistic frontal cortex, motor, and somatosensory areas. And it continues to enhance and work on visual and auditory cortex areas and other important structures such as the hippocampus, basal ganglia, and thalamic systems. CCortex Developer Box includes a 64 bit Spiking Neural Network Engine. Each Developer Box can represent up to 250 million neurons containing 11 000 synapses, and will be capable of updating a data matrix of 1.5 billion synapses 10 times per second for real-time applications. (Source: http://ad.com/).
3. From the Blue Brain project, Henry projected that Blue Column (simulating 100 000 neurons) will be possible by about 2008/9, and the detailed cellular level model within five years, while the model at the cellular level of the entire brain will probably take a decade.
4. Soar is one of the most established cognitive architectures with a wide range of military applications. The new research focus of Soar 9 includes extensions of Soar cognitive architecture and the integration of emotions with reasoning and learning mechanisms. More information is available at http://sitemaker.umich.edu/soar/home.
5. Detailed information for ACT-R can be found at http://act-r.psy.cmu.edu/about/.
6. ICARUS operates on a recognize-act cycle and it focuses on reactive execution of existing skills rather than a problem-space search. It generates reactive behaviors to achieve goals through constant

interaction with the environment. ICARUS lacks planning capability in this sense as it constantly has to wait for percepts from the environment. See also http://www.isle.org/langley/talks/icarus.6.04.ppt.

7. *DARPA Quest.* In the DARPA call for BICA (Biologically-Inspired Cognitive Architecture) phase one BAA, we see a number of brain-inspired designed architectures. A very brief summary is provided here:

   a. *CHIP: A Cognitive Architecture for Comprehensive Human Intelligence and Performance* (Shrobe *et al.*, 2006).

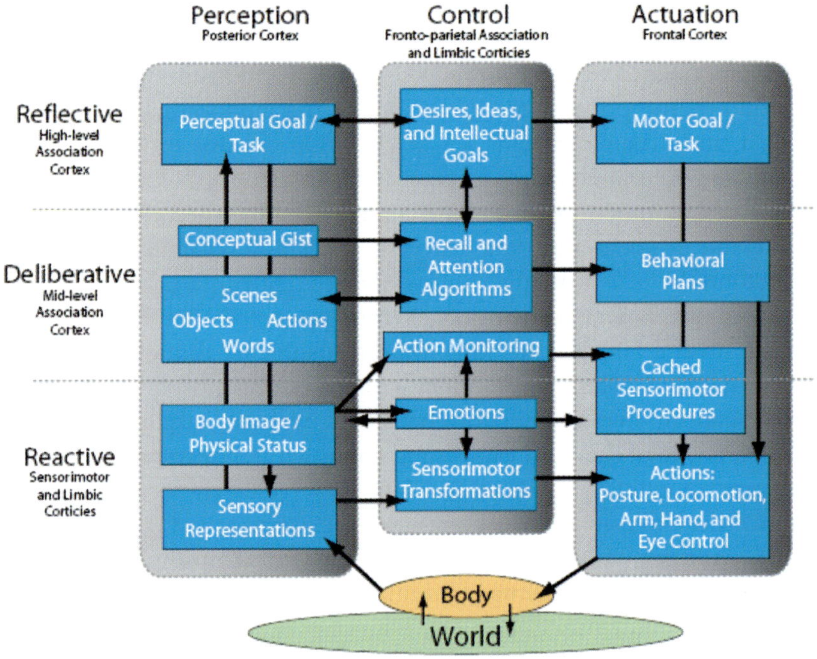

The figure shows the CHIP Architecture: the systems of components intimately interact with one another, forming tightly integrated loops through the various layers. The middle column, which manages the internal state of the system, acts as a shared blackboard with three levels for general communication among the components. But in addition to this shared channel, there are many special purpose channels linking specific sub-systems (Shrobe *et al.*, 2006), MIT. This is a well thought architecture that includes the Sloman's three-layer architecture and the brain's cortex arrangement).

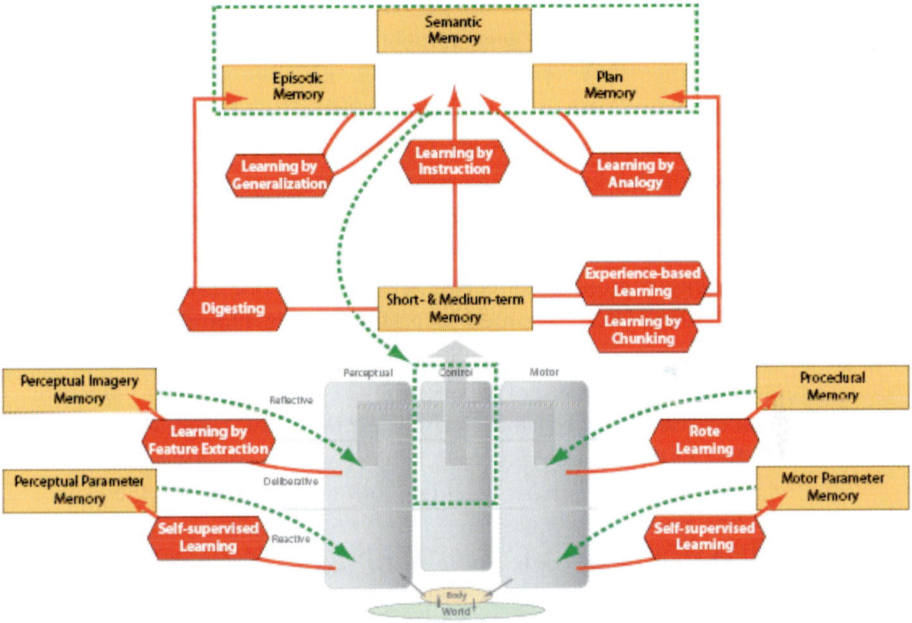

The figure shows the memory systems and corresponding learning pathways in the CHIP architecture.

b. *TOSCA Architecture Design*

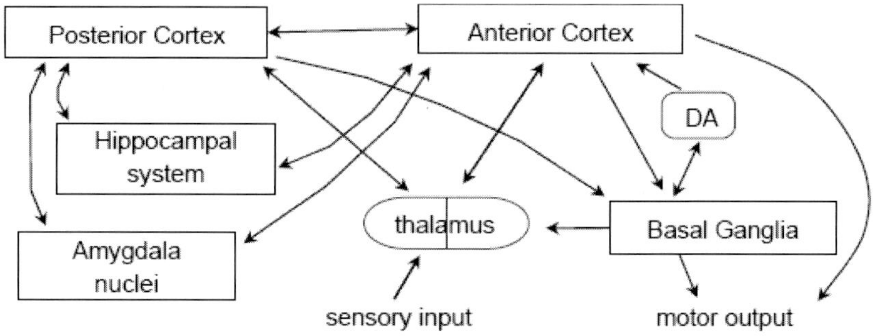

The figure shows the major pathways within and between neural systems underlying TOSCA architecture design (by Professor John Laird, University of Michigan).

Professor John Laird's key research focus is on the Soar cognitive architecture discussed in the chapter. TOSCA is another biological-inspired architecture Laird works on, and more lately, TOSCA has been renamed as STORM (see Douglas Pearson's website from Three Penny Software). STORM's key focus is to create a framework that is easy for the architecture to work on such as the ability to change the module easily.

In TOSCA's architecture, the pattern of connections is claimed to be consistent with the following processing loop:

(1) The posterior cortex potentially receives five types of input: sensory input (via the thalamus), episodic memory (via the hippocampal system), top-down control signals (from the anterior cortex), emotion (amygdale), and bottom-up associations (from other parts of the posterior cortex).
(2) Specific (core) thalamocortical loops cluster on patterns of activity, recognizing familiar input patterns.
(3) Non-specific (matrix) thalamocortical loops encode (and retrieve) sequences of clusters, producing a representation of what is expected to come next.
(4) Cortico-cortical connections within the posterior cortex modify and elaborate the internal state, generating a more

complete hierarchy of clusters of sequences of clusters that is organized topographically within cortical areas.

(5) Cortico-cortical projections to the anterior cortex propose specific intentions, which could be motor actions or mental actions (e.g. setting cues for episodic memory retrieval, setting goals in working memory, maintaining or attending to specific information).

(6) Corticostriatal loops select among competing actions based on the (learned) values associated with each action in the current context/state. Multiple actions can be selected in parallel, based on the parallel structure of frontostriatal loops.

(7) State-action values are modified by the midbrain dopamine system in a way that realizes reinforcement learning algorithms (strengthening state-action associations that lead to long-term rewards).

(8) Motor actions are passed on to the motor output systems for execution whereas mental actions provide top-down control signals to the posterior cortex.

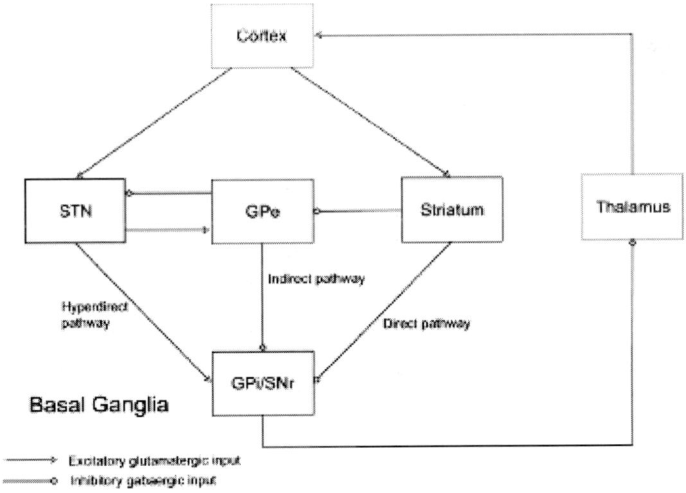

The figure shows the schematic of corticostriatal pathways.

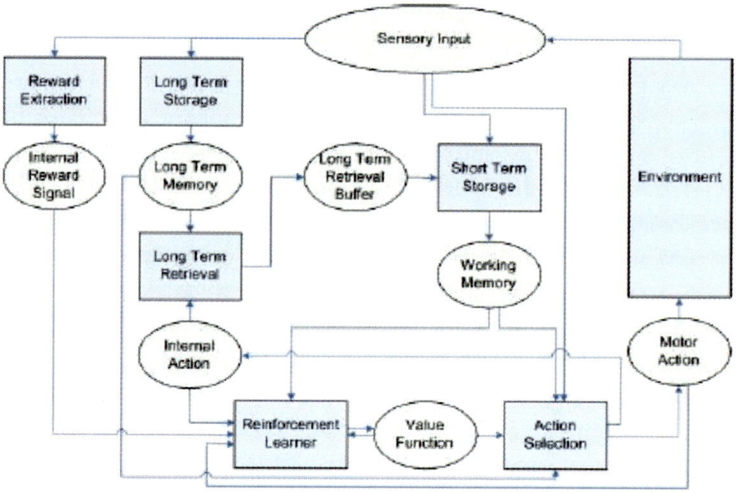

The figure shows an initial STORM framework. (Source: http://www.eecs.umich.edu/~soar/sitemaker/workshop/27/Gorski2.pdf).

c. *Self-aware cognitive architecture*. This is another architecture proposed by GMU for BICA phase one. The name of this architecture is due to its main feature, which is the notion of self-aware cognition. Again, it has the same claim as inspired by studies of the human brain-mind: in particular, theoretical models of representations of agency in the higher associative human brain areas. The self-aware cognition feature is said to allow the system to maintain human-like attention, focus on the most relevant features and aspects of a situation, and come up with ideas and initiatives that may not follow formal logic (DeJong *et al.*, 2006).

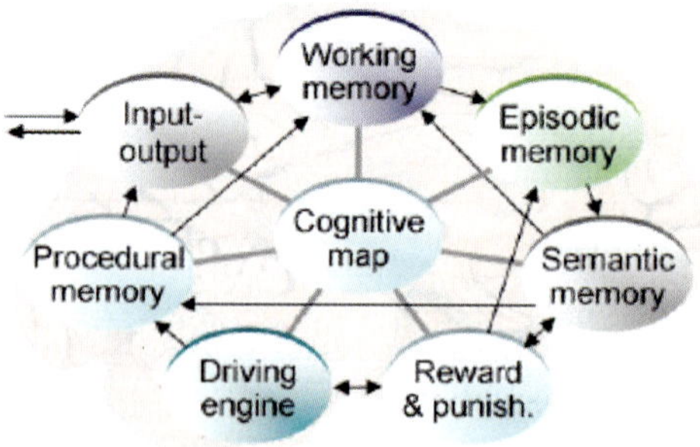

The figure shows the top level self-aware architecture. It is based on three key building blocks, namely schemas, mental states, and cognitive maps. The schemas are used for representation of knowledge and experiences. Mental states are used for instantiating a self. And the cognitive maps are to provide efficient indexing and navigating of stored memories.

d. *SAL Architecture.* SAL stands for Synthesis of ACT-R and Leabra. SAL brings together ACT-R, with Leabra (Leabra is a high neural-fidelity architecture that is claimed to perform well in adapting to new environments).

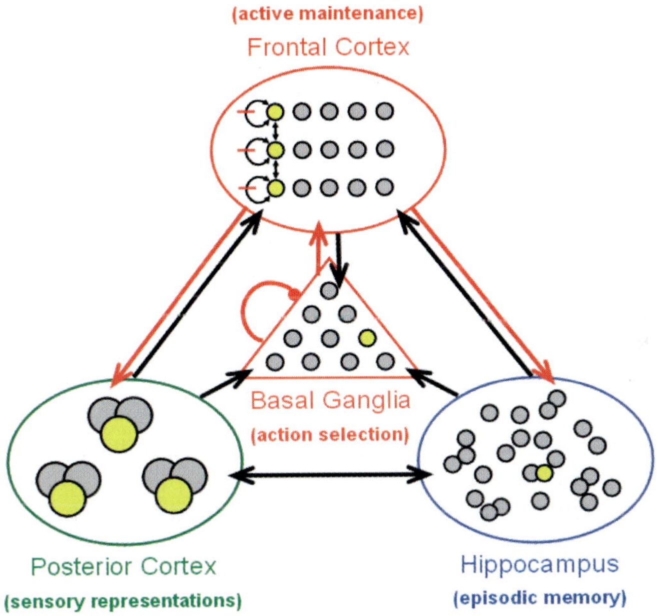

The figure shows the tripartite architecture of SAL. Human cognition is conceptualized in terms of the computational properties of distinct brain areas, each specialized for different incompatible forms of learning (e.g. rapid learning in the hippocampus versus slow learning in the cortex). Red arrows represent top-down cognitive control (which results from interactions between the frontal cortex and basal ganglia), while black arrows represent standard neural communication.

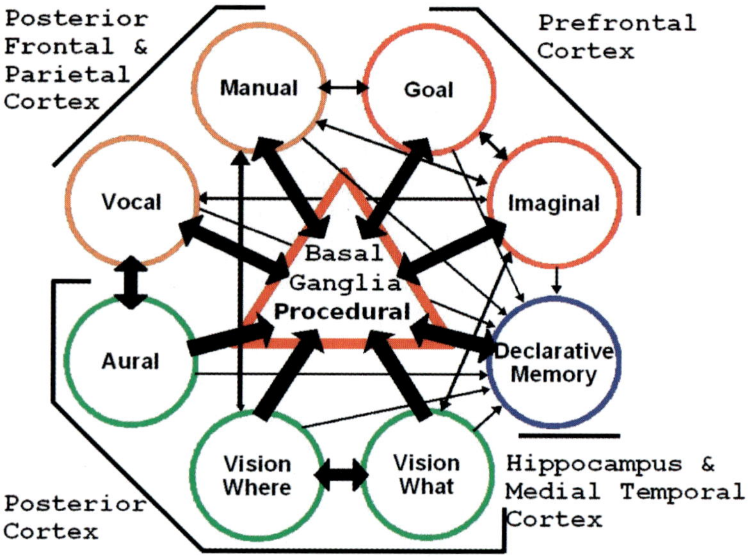

The figure shows the cognitive module architecture of SAL, where the broad tripartite architecture has been subdivided into finer grained separable cognitive mechanisms.

e. *BICA-LEAP Architecture*. BICA-LEAP stands for Biologically-Inspired Cognitive Architecture for integrated LEarning, Action, and Perception. BICA-LEAP is a joint proposal from HRL Laboratories, University of Southern California (USC), Boston Univeristy CNS department (BU/CNS), and Portland State University (PSU). BICA-LEAP attempted to synthesize a single comprehensive architecture based on core brain operating principles and computational paradigms that can be adapted to solve all the modal problems, visual object recognition, audition, motivation, etc). Till today, these modal problems are solved individually, i.e. in disparate architectures whose design embodies specializations for each model problem (from the talk by Dr Deepak Khosla, HRL Laboratories, in the BU/CNS ICCNS 2007 conference).

Chapter 11

1. Plastic Electronics — The building of plastic electronics may revolutionize the computer technology into the next "S" curve. In plastic electronics, electronic components can be assembled in thin plastic form. The facility will produce flexible active-matrix display modules for "take anywhere, read anywhere" electronic reader products. It will utilize Plastic Logic's unique process to fabricate active-matrix displays that are thin, light, and robust, enabling a reading experience closer to paper than any other technology. (Press release FT 3 Jan 2006).
2. Reliance on the computer system is greater than ever. We will have a "system log me out" phenomenon, and without the computer system, it could not work.
3. People's desire for knowledge will be much greater and information exchange will be great — across the internet and travel.
4. Multimedia technology will have a great impact, keeping many people busy and making them lose touch with real life.
5. Communication technology is going to do wonder, and many exciting stuff will be marketed. Communications media has started from humble paper letter writing (posted by pigeons) and progressed to long distance telephone connections, email, internet chat rooms, video conferences, and 3-D virtual representations of our interacting environment.
6. Mind chip. Many will still attempt to find out how the brain works and want to model it to develop a mind chip. (See also New Scientist magazine, dated 3 February 2007, about the mind chip.)

Chapter 12

1. Many scientists indicate the brain has an associative network connection. References include (Gros, 2005).
2. One possible explanation for the diverse view of how the brain works is that the brain is a complex machinery and everyone is looking at a spotlight of the big picture.

3. The technique for deconstructing brain wiring at the single neuron level is still in the infancy stage.
4. Some information on the localization issues of the brain are the recorded notes from a talk by Professor Rebacca Saxe (MIT/BCS) in BU in the February 2007 Neuphi meeting.
5. Open issues include:

   a. The motor system, such as how the projection is done well in advance.
   b. The association cortex and how it works. There remain many research issues at the association cortex.
   c. How are the detailed cortical connections wired?
   d. Why is the cerebral cortex arranged in six layers, why does the cortical column have patterns such as a pinwheel pattern, and why is the brain structure hierarchical?
   e. At what scale is the architecture of the mammalian brain organized? Many people believe this reduces to both an empirical question in neuroscience and a theoretical question in evolutionary biology. Evolutionary biology asks whether there is reason to believe that the brain evolved modules through the mechanism of natural selection, and whether these modules can be thought of as achieving definable goals. (Glimcher, 2004).
   f. Why do we have this inborn moral law? The ability to discern between right and wrong seems hard-wired in the brain. Why is this process built-in?
   g. Why does the brain have rhythms? How are the different brain waves related to our thinking mechanism and brain states? Why are these rhythms related to the respective function? For example, why are gamma rhythms related to attention and why is the delta function at rest state? Where do these rhythms come from? What are they good for? How do they work together in our brain?
   h. Why do the neural activities keep changing even in the presence of nothing-new-to-learn situations?
   i. How many of the processes are innate (nature) or developed (nurture)? Innate or developed (derived)? It is also not clear how much the modular mechanisms are innate and how much is the process of canalization and increase of specialization with development (where canalization is the development of an organism

along relatively predictable pathways despite abnormality or injury). For example, "words" are non-innate specialization. It appears that specialization regions developed to handle words are clearly not innate. Left fusiform gyrus are recruited in looking for words in our own language rather than in any other language. What about the RTPJ (right temporo-parietal junction) to belief attribution?

j. Why do we have déjà vu experiences? Why do we have déjà vu experiences? The term "déjà vu" is a French word for "already seen". It is an experience where one feels that an event or person is something he has seen before or has a sense of familiarity with, and it also comes with a sense of "eeriness" or "strangeness". Studies suggest that the déjà vu experience is common to most people.
k. Why do we have dreams? Is dreaming essential for learning? Does dreaming help in sorting information in the brain? Do dreams have anything to do with our survival strategies or issues? Jonathan Winson, a psychiatrist, suggested that all long-term memories may be constructed through this off-line process during REM sleep. During sleep, the hippocampus would process the day's events and store important information in the long-term memory. Winson suggested that the dream is an ordered processing of memory, which interprets experience that is precious for survival. Dreams are essential to learning. (Scaruffi, 1999).
l. Why is sleep so important for the function of the brain?
m. Why is there this unknown dark energy? Neuroscientists are puzzled by the additional energy required for the brain to perform mental tasks, which is extremely small compared with the energy that the brain expends as an individual who does nothing at all. (Studies of brain activity are mostly from fMRI). Why does the resting brain generate so much energy? (This is also known as dark energy (Raichle, 2006).) Raichle proposed three possible explanations for the brain's dark energy:

    i. People have random thoughts and daydreams when doing nothing.
    ii. Intrinsic activity might emerge from neural efforts to balance the opposing signals of cells simultaneously trying to jack up and cool down brain activity.

iii. Internal process of generating predictions about upcoming environmental demands and how to respond to them.

n. Why do some autistic people have special abilities in using their brain? For example, Peek Kim claimed to be able to remember all the facts he read and many others claim to have special artistic skills such as illustrating city landscapes in great detail.

6. Can we reverse engineer the brain using computational techniques? To start with, the brain and the human body are not a computational system per se. It is a biological and chemical system. We know some aspect of how the brain works, but we may not know how to reproduce or model precisely and accurately the biological processes. And we also cannot explain what goes on inside the brain and why certain biological processes take place as desired.

   Could we then reverse engineer to model the brain using computational techniques? This is still a debatable issue.
7. Difficulty in tracing the "wires" of the brain, its axons, and dendrites. Seung's lab/MIT in collaboration with Winfied Denk (Max Planck Institute — Heidelberg), who invented a new technique called serial block-face scanning electron microscopy. This automated technique claims to yield 3-D images of the brain at nanoscale resolution. For more information, see http://hebb.mit.edu/index.html.
8. The visual system has been most well studied, yet there still remain some critical unsolved issues. These include:

   a. Human LGN. Human LGN has six layers in the macular region, which fuses into four in the periphery and leaves two in the monocular zone. Occasionally, there is also an eight-layer area at the edge of the four-layer segment (Hickey and Guillery, 1979). The functional significance of this has not been investigated.
   b. How does the brain analyze surface appearance, such as color, texture, and gloss? These are important and critical for our everyday tasks such as deciding whether a patch of pavement is icy, whether a pancake is cooked, and whether a road is muddy before stepping on it. (Professor Edward Adelson/MIT, his graduate student Lavanya Sharan).

c. Which are the neural mechanisms for detecting image skewness? We know that in the retina and brain, there are cells that are preferentially sensitive to either bright patterns or dark ones. This will help scientists to develop better visual systems for robots so that the robot can make better judgments based on surface appearance.

9. *Neocortical design issue.* What are the transforming operations imposed in a local region of the neocortex, a cortical column, upon its input to produce several outputs? The essays included here provide a cross-section of this large field, written by investigators using methods that include Golgi studies, slice recordings with multiple intracellular microelectrodes and multiple microelectrode recordings in the intact cortex; several include theoretical modeling. While no one of these authors would venture to claim the problem is solved, their contributions and those of others in the field indicate that significant progress has been made in constructing an intra-columnar flow diagram, and in understanding the dynamic neuronal operations within it. (Mountcastle, 2003).
10. *Prediction and estimation.* We have amazing prediction and estimation capabilities. If you close your eyes for a short duration, you will still have the perception to move for a short distance or the ability to point your finger to a person around you without much difficulty. The brain's ability to visualize and perceive in predicting and estimating the physical world is another beautiful design that we have not fully grasped. What is the mechanism behind this capability? That is the ability to predict and estimate from static or moving occlusion to a more natural scenario such as playing a game like tennis. In playing tennis, you will notice that your arm and motion control are able to move to the correct placement way before the ball returns to your court. Our ability to constantly make prediction and estimation in our motor movement and thinking process is still a wonder (Ooi, He, Wu, 2001).
11. *Brain flattening method.* The surface of the human cerebral cortex is a highly folded sheet with the majority of its surface area buried within folds. As such, it is a difficult domain for computational as well as visualization purposes. A number of studies on brain flattening (Schwartz *et al.*, 1989) (Wolfson and Schwartz, 1989) inflate it so

that activity buried inside sulci may be visualized (Fischl *et al.*, 1999) and segmentation of the brain's gray matter, white matter, skull dura, pia, etc (Fischl *et al.*, 2002). As said, reconstruction is a complex task that requires the solution of a number of subtasks such as intensity normalization, skull-stripping, filtering, segmentation and surface deformation (Dale *et al.*, 1999), and segmentation at sub-cortical regions is a major problem and challenge.

12. *Brain rhythms.* We know that the brain generates the following rhythms (list not complete):

    a. Delta (1–3 Hz) deep sleep. If awakened, one acts groggy and confused (observed in very deep sleep to the resting phase).
    b. Theta (4–7 Hz) is observed in the learning and memory process. In (O'Keefe and Recce, 1993), the spatial information is coded at theta frequencies by clusters of neurons segmentally distributed along the longitudinal axis of the hippocampus) — in the wakefulness stage.
    c. Alpha (8–12 Hz) involves deep relaxation or just an awake and drowsy state, often with the eyes closed. It is believed in alpha, we begin to access the wealth of creativity that lies just below our conscious awareness. Some believe it is the gateway to the entry-point that leads into deeper states of consciousness. (A recent study links it to modulating selective attention. Alpha power, observed in highly trained animals, is modulated as a function of cued attention and can show column-specific expression to suppress input from a "distractor" and can show a column-specific expression to suppress input from distractor vibraissae.
    d. Beta (12–30 Hz) is an alert, attentive, and thinking stage involving rapid firing, heightened alertness, and visual acuity. It may be key to the act of cognition. Novelty (?) Long range?
    e. Gamma (30–90 Hz) is an attentive state. It is present when we are awake and paying attention. It is reduced during sleep and disappears when we are totally unconscious, such as when under anesthesia. These cognitive activities have been associated with gamma (40 Hz in particular) — attention, memory, facial recognition, REM sleep, and the comprehension of new concepts. Gamma is thought to be important to binding (the ability of the brain to combine all inputs (external or internal)

into a single unified conscious experience) (from http://www.mindupdate.com).

13. *Brain rhythms.* "There is a great deal of experimental evidence that rhythms in the brain play a significant role in perception and cognition." Mike Denham and Miles Whittington in Cognitive Systems Project InterAction Conference (IAC, 2003). They indicated that future computational architectures for cognitive systems will involve spatially distributed, functionally-specialized information processing regions, like those in the brain, and will similarly require mechanisms for coordinating activity across these distributed processes. The synchronization of rhythmic activity between distributed processes may be a candidate mechanism to enable the efficient operation of such architectures. Denham and Whittington also listed the following key questions and emphasized the importance to understand the underlining mechanisms:

    a. What are the neural mechanisms of rhythmic activity and synchronisation, both locally and across widely distributed regions?
    b. Why the interplay between slow and fast brain rhythms?
    c. What are the roles of synchronisation over different frequency bands?
    d. How does the brain build a coherent perceptual account of a sensory event in the case that the component features of the event are processed asynchronously in widely distributed areas of the cortex?
    e. Is rhythmic activity fundamental to this process?
    f. Will future artificial sensory systems require similar mechanisms?
    g. Does rhythmic activity play a role in the organization, storage, and retrieval of episodic memories, and if so, what role is it, e.g. "chunking" of individual perceptual/cognitive experiences into a complete "episode"?
    h. Does this have any impact on the way information is composed of sequences of events that might be stored/retrieved in future artificial cognitive systems?
    i. If future artificial cognitive systems employ "massively" distributed asynchronous processing hardware architectures, will they face the same problems as the brain in providing coherent behavior?

j. Is rhythmic, synchronized activity in the brain dependent on intrinsic neural mechanisms or is it an "emergent" behavior of the brain resulting from inherent self-organizing, adaptive processes?
k. If so, would we expect to observe it as an emergent feature of any massively parallel, distributed self-organizing computational architecture when it is required to deliver coherent behavior?

14. *The neuron as a functional unit process input verses as an oscillator, resonating and modulating endogenous activity.* Hebb said, "Cell that fire together wired together". If Cell A fires before Cell B fires, the connection link is strengthened, and if Cell A fires after Cell B fires, then the connection link is weakened. But this is not true if both Cell A and Cell B are oscillators. (Reference to Nancy Kopell's research work, Math department, BU). Nancy Kopell and her research groups in BU have shown several computational models based on neural rhythms and oscillation to demonstrate how spiking in a neuronal network can be brought about by the interaction between oscillating excitatory cells and inhibitory cells (Borgers and Kopell, 2004), how the cortical network supports attention (Borgers *et al.*, 2005), and how the gamma and theta oscillations are observed within the hippocampus for learning and memory (Gloveli *et al.*, 2005). For more information, visit: http://math.bu.edu/people/nk/.
15. These are grey neuroscience areas but interesting to know:

a. Belief attribution. The representation of belief, thought, and any internal states seem to be part of the RTPJ areas in the brain (Saxe and Sch, 2003).
b. The motor cortex can also do a kind of thinking. Georgopoulos of the University of Minnesota reported that the motor cortex can help to recognize and remember the sequence of events in time. Georgopoulos's team found that neuronal activity in the motor cortex can provide information about the direction of an upcoming movement and can also serve as a memory for the spatial locations of individual stimuli to which the animal was supposed to move. This potentially means that the information needed to perform complex cognitive tasks is distributed widely in the brain (Wickelgren, 1999).

Chapter 13

1. What is the limitation of the human brain? Listed below are some possible limitations of the brain and how the computer can be used to augment these limitations:

   a. The number of synapses and neurons are more or less fixed in the adult brain. The brain's "computational power" is computed to be around 100 teraflops and 100 terabytes (see Chapter 1). The number of synapses and neurons do not continue to increase but stay more or less constant throughout our life after the early development period is over.

      i. The computer can be used to store huge information that may be useful in the future. (As storage is getting cheaper, do we need to remove anything completely from memory? How do we decide what to forget? Do computers dream? Does forgetting involve removal from memory or does it make memories harder to find (foreground and background info, removing links)?)
      ii. Design devices that could help people with memory difficulty.

   b. Memory and remembering issue. Our memory is not totally accurate and subject to illusory or imaginary influence. Seeing a protruded hollow face is an example of illusion that can fool us. Our memories are also easily influenced by top-down information (although top-down information does help us in a lot of areas as explained in the earlier chapter). Our memory likes to perceive and believe what we want to believe even at times when it is at fault. As we grow older, we are in general also more "stubborn" and harder to change. We start to have self-pride. And depending on the type of cultural influence, we may have a certain attitude that makes our perceived information sometimes difficult to believe, even of ourself. In layman terms, we call it a "decease" in our brain caused by fear, stubborn, etc. Why is this so in the human brain?
   c. We do not remember all the information that we see every day. But we remember information that looks important to us and

we pay attention to it. This is good as we do not unnecessarily load our memory system. However, we may miss out essential information.

d. Physical fatigue. Humans are physical beings and subject to fatigue and stress. Machines would not have the fatigue.
e. Humans need to activate or exercise both the brain and body. Without that we will die. But machines do not need this.
f. Spatial updating. Spatial updating refers to the cognitive process that computed the relationship between oneself and the entire world changes. Experiments have shown that we update local spatial relationships but do not update the global relationship. However, if the participant automatically updated the local targets when they moved relative to the global environment (e.g. relative to the campus), the global environment is updated, as well as the local environment (Wang and Brockmole, 2003).
g. The human brain can only maintain a limited number of different thoughts before it starts to get confused.
h. Humans may have negative emotions such as depression.
i. Humans remove complexity by ignoring details. But if machines have unlimited resources or more resources than humans (beyond what humans can do), then they can still attend to more things in the same time span.
j. Humans can be distracted, but machines may be designed to consider all factors, e.g. machines can have four eyes (so even if there is one distractor in one eye, the rest can still function normally). (Still, a question of unlimited resources versus attention?)
k. Human intelligence can take time to build, but be very robust; a baby takes two years to say the first word (if we were to build intelligent machines, we cannot afford to wait for it to learn for over 20 years before it amasses enough knowledge).
l. The brain does make mistakes. We experience that in our life.

2. Psychologists have long known that our memories are easily embellished. We add imaginary details through wishful thinking or to make a more logical story that sometimes is not the true story.

Chapter 14

1. Understanding the autistic mind. Interesting poems from Tito, an autistic child, can be found in http://magma.nationalgeographic.com/ngm/0503/feature1/online_extra.html.
2. It has been demonstrated that children with autism typically have special difficulties in understanding the beliefs of others. This suggests that they lacked the "theory of mind" (Williams *et al.*, 2001).
3. The three laws of computers (or robots) (Picard, 1997):
   a. A computer may not injure a human being, or through inaction, allow a human being to come to harm.
   b. A computer must obey the orders given it by human beings except where such orders would conflict with the first law.
   c. A computer must protect its own existence as long as such protection does not conflict with the first or second law.

   — Reproduced from the "The Bicentennial Man" by Isaac Asimov. His robots are subject to "The Three Laws of Robotics", reproduced by Picard, except the word "robot" is replaced by the word "computer".
4. False belief test: The Sally-Ann test, which depicts the behavior of two little girls, Sally and Ann. In the first frame, Sally places a doll in her carriage. When she leaves the room, Ann takes the doll and places it in her box. In a moment, Sally will return to the room. Now, please answer the following three questions:

   Where is the doll?
   Where did Sally put the doll?
   Where will Sally look for the doll?

   Sally will look for the doll in her carriage because that is where she mistakenly believes it is; a child of three, however, will answer that Sally will look in the box because that is where the doll actually is. However, the test isn't about location of the doll, but about what Sally is likely to believe in regard to its location. Three year olds fail the test because, in order to come up with the correct answer, they have to mentally put themselves in Sally's place, and they cannot yet do that.

Neuroscientists use the Sally-Ann test and other false-belief (FB) tests to evaluate one person's ability to attribute mental states to another person. This is a gradually developing process that occurs unevenly within the population. By age six to seven, the vast majority of us are able to pass much more challenging FB tests by entering into another person's mind by proxy and making an accurate approximation of how that person is thinking or feeling under particular circumstances. It has been observed under neuroimaging that children aged six to seven years old have used the cortical regions such as the bilateral temporo-parietal junction (TPJ), posterior cingulate, and medial prefrontal cortex, which are responsible for TOM. In adults, the right TPJ is recruited selectively when subjects read about a character's thoughts, but not other facts about a person. Developmental change in response selectivity was observed in just this one region: the right TPJ was recruited equally for any information about people in younger children, but only for descriptions of mental states in older children (Saxe and Wexler, 2005).

5. *Mind opening questions.* Here I would like to discuss some mind opening questions (or to some people it could be "mind boring" questions).

   a. Why are we 99.9% genetically identical (Collins, 2006; all human kind have a surprisingly low level of genetic diversity compared with other species) but yet also so different and unique in many ways? Even a twin is different. We (human beings) look so much alike in many areas, but yet we are still so different as an individual. For example, our fingerprints are all different, and our eye retina patterns are all different. There is no known identical copy.
   b. Why do the atoms and molecules hold together? What is this force that holds them together? What holds our body together? I mean, why does such a force exist in the first place? If that force is released, we will all be disintegrated and vaporized. Imagine, if the gravitational force disappears from Earth, we will suddenly be floating away from Earth into space.
   c. Why do we have an inborn moral law in us? Irrespective of any race and place you are born, you can from birth distinguish

what is right and wrong. Children do not need to be taught; they immediately can distinguish what is right and what is wrong in behavior. More surprising, most parents will know that little kids do not need to be taught to do wrong things and behave badly.

d. Why do we have so many distinct languages? For example, there are striking differences between Russian, Chinese, and Indian languages, even though the people who speak these languages are geographic neighbors and have pretty much the same brain.

REFERENCES

Abrams M. (2003). Can you see with your tongue? *Discover* **24**(6).

Ahissar M, Hochstein S. (1993) Attentional control of early perceptual learning, *Proc Natl Acad Sci USA* **90**: 5718–5722.

Ahissar M, Hochstein S. (1998) Perceptual learning. In: Walsh & Kulikowski J (eds) *Perceptual Constancies: Why Things Look As They Do*, pp. 455–498. Cambridge University Press.

Ahissar M, Hochstein S. (2004) The reverse hierarchy theory of visual perceptual learning. *Trends Cog Sci* **8**(10).

Albrecht DG, Geisler WS. (1991) Motion selectivity and the contrast-response function of simple cells in the visual cortex. *Vis Neurosci* 7: 546.

Alexander SPH, Peters JA. (2000) Receptor and ion channel nomenclature supplement. *Trends Pharma Sci.*

Allman JM, Kaas JH. (1974) The organization of the second visual area (VII) in the owl monkey: A second order transformation of the visual hemifield. *Brain Res* **76**: 247–265.

Alvarez GA, Cavanagh P. (2005) Independent Resources for Attentional Tracking in the Left and Right Visual Hemifields. *Psych Sci* **16**(8): 637–642.

Amir E, Maynard-Zhang P. (2004) Logic-based subsumption architecture. *Artificial Intelligence* **153**: 167–237.

Ananthanarayanan R, Modha DS. (2007) Scaling, stability, and synchronization in mouse-sized (and larger) cortical simulations. *CNS*2007:*

Sixteenth Annual Computational Neuroscience Meeting, Toronto, Canada July 8–12 2007. (Also appears as *IBM Research Report RJ 10405, 2/14/2007.*)

Anderson J. (1993) *Rules of Mind.* Lawrence Erlbaum Associates, Inc.

Anderson J. (2003) Arithmetic on a parallel computer: Perception versus logic *Brain and Mind* **4**: 169–188.

Anderson JR, Bothell D, Bryne MD, Douglass S, Lebiere C, Qin Y. (2004). An integrated theory of the mind. *Psych Rev* **111(4)**: 1036–1060.

Anderson JR. (2005) Human symbol manipulation within an integrated cognitive architecture. *Cog Sci* **29**: 313–341.

Andersson Russell L. (1988) *A Robot Ping-Pong Player: Experiment in Real-Time Intelligent Control*, ISBN: 9780262011013, MIT Press.

Ashby F, Alfonso L, Turken A, Waldron E. (1998) *Psych Rev* **105**: 442.

Baddeley AD, Gathercole SE, Papagno C. (1998) The phonological loop as a language learning device. *Psych Rev* **105**: 158–173.

Baddeley AD. (2000) The episodic buffer: A new component of working memory? *Trends Cog Sci* **4**: 417–423.

Baddeley AD, Wilson BA. (2002) Prose recall and amnesia: Implications for the structure of working memory. *Neuropsychologia* **40**: 1737–1743, 22.

Barbas H. (1995) Anatomic basis of cognitive-emotional interactions in the primate prefrontal cortex. *Neurosci Biobehav Rev* **19**(3): 499–510.

Bar-On R, Tranel D, Denburg NL, Bechara A. (2003) Exploring the neurological substrate of emotional and social intelligence. *Brain* **126**: 1790–1800.

Basser PJ, Pajevic S, Pierpaoli C, Duda J, Aldroubi A. (2000) In vivo fiber tractography using DT-MRI data. *Magn Reson Med* **44**: 625–632.

Bechara A, Damasio H, Dsmasio AR, Lee GP. (1999) Different Contributions of the Human Amygdala and Ventromedial Prefrontal Cortex to Decision-Making. *J Neurosci* **19**(13): 5473–5481.

Bechara A, Damasio H, Damasio AR. (2003) Role of the amydala in decision-Making. *Annals New York Acad Sci* **985**: 356–369.

Besl PJ, Jain RC. (1986) Invariant surface characteristics for 3D object recognition in range images. *CVGIP* **33**: 33–80.

Bhatt R, Carpenter G, Grossberg S. (2007) Texture segregation by visual cortex: Perceptual grouping, attention and learning. *Technical Report CAS/CNS-TR-2006–2007*, Boston University. In Press, Vision Research.

Synaptic modifications in cultured hippocampal neurons: Dependence on spike timing, synaptic strength, and postsynaptic cell type. *J Neurosci* **18**(24): 10464–10472.

Bickhard MH, Terveen L. (1995) *Foundational Issues in Artificial Intelligence and Cognitive Science — Impasse and Solution.* Elsevier Science Publishers.

Bless H, Schwartz N, Kemmelmeier M. (1996) Mood and stereotyping: The impact of moods on the use of general knowledge structures. *Eur Rev Social Psych* 7: 63–93.

Bliss TV, Collingridge GL. (1993) A synaptic model of memory: Long-term potentiation in the hippocampus. *Nature* **361**(6407): 31–9.

Bliss TV, Lomo T. (1973) Long-lasting potentiation of synaptic transmission in the dentate area of the anaesthetized rabbit following stimulation of the perforant path. *J Physiol (London)* **232**: 331–356.

Borg-Graham LJ, Monier C, Fregnac Y. (1998) Visual input evokes transient and strong shunting inhibition in visual cortical neurons. *Nature* **393**(6683): 369–373.

Borgers C, Epstein S, Kopell NJ. (2005) Background gamma rhythmicity and attention in cortical local circuits: A computational study. **102**(19).

Borgers C, Kopell N. (2004) Effects of noisy drive on rhythms in networks of excitatory and inhibitory neurons. *Letter in Communicated by Bard Ermentrout.*

Born RT, Bradley DC. (2005) Structure and Function of Visual Area MT. *ARI* **16 March 2005**: 157–189.

Bower GH. (1991) Emotional mood and memory. *Amer Psych* **31**: 129–148.

Brooks R. (1999) *Cambrian Intelligence: The Early History of the New AI.* MIT Press.

Brooks Rodney. (2003) *Flesh and Machine, reprinted.* Vintage.

Brown J, Bullock D, Grossberg S. (1999) How the basal ganglia use parallel excitatory and inhibitory learning pathways to selectively respond to unexpected rewarding cues. *J Neurosci* **19**: 10502–10511.

Brown JW, Bullock D, Grossberg S. (2004) How laminar frontal cortex and basal ganglia circuits interact to control planned and reactive saccades. *Neural Networks* **17**: 471–510.

Bradski G, Carpenter GA, Grossberg S. (1994) Store working memory networks for storage and recall of arbitrary temporal sequences. *Biol Cybernetics* **71**: 469–480, Springer-Verlag.

Bratman ME, Israel DJ, Pollack ME. (1988) Plans and resource-bounded practical reasoning. *Comput Intelligence* **4**(4): 349–355.

Bullier J. (2001) Integrated model of visual processing. *Brain Res Rev.*

Bulkin DA, Groh JM. (2006) Seeing sounds: Visual and auditory interactions in the brain. *ScienceDirect, Curr Opin Neurobiol* **16**: 415–419.

Bunge SA, Wendelken C, Badre D, Wagner AD. (2005) Analogical reasoning and prefrontal cortex: Evidence for Spearable retrieval and integration mechanisms. *Cerebral Cortex* **15**(3): 239–249.

Butler G, Gantchev A, Grogona P. (2001) Object-oriented design of the subsumption architecture. *Software — Practice Experience* **31**: 911–923.

Carandini M, Heeger DJ. (1994) Summation and division by neurons in visual cortex. *Science* **264**: 1336.

Camerer CF. (2003) *Behavioral Game Theory: Experiments in Strategic Interaction.* Princeton: Princeton UP.

Carey S. (1985) *Conceptual Change in Childhood.* Cambridge, MA, MIT Press.

Carey B. (2007) Subconscious runs with external cues. *The New York Times, July 31, 2007.*

Carpenter GA, Grossberg S. (1987) A massively parallel architecture for a self-organizing neural pattern recognition machine. *Comp Vision Graphics Image Processing* **37**: 54–115.

Carpenter GA, Grossberg S, Rosen DB. (1991) Fuzzy ART: Fast stable learning and categorization of analog patterns by an adaptive resonance system. *Neural Networks* **4**: 759–771.

Carpenter GA, Grossberg S, Markuzon N, Reynolds JH, Rosen DB. (1992) Fuzzy ARTMAP: A neural network architecture for incremental supervised learning of analog multidimensional maps. *IEEE Transactions on Neural Networks* **3**: 698–713.

Carpenter GA. (1997) Distributed learning, recognition and prediction by ART and ARTMAP neural networks. *Neural Networks* **10**: 1473–1494.

Carpenter GA, Grossberg S. (2003) Adaptive resonance theory, In: Arbib MA (Ed), *The Handbook of Brain Theory and Neural Networks, 2nd ed.*, pp. 87–90. Cambridge MA: MIT Press.

Carruthers, Peter (1996) Simulation and self-knowledge: A defence of theory-theory. In: Carruthers Peter and Smith Peter K (eds.), *Theories of Theories of Mind*, Ch 3, pp. 22–38. Cambridge University Press.

Carruthers P, Smith PK. (1996) *Theories of Theories of Mind.* Cambridge University Press.

Chalmers DJ. A Computational Foundation for Study of Cognition, ID code 319, Deposited in Cogprint on 15 June 1998.

Chater N, Tenenbaum JB, Yuille A. (2006) Probabilistic models of cognition: Conceptual foundations. **10**(7): 287–291.

Chater N, Tenenbaum JB, Yuille A. (2006) Probabilistic models of cognition: Where next? *Trends Cog Sci* **10**(7): 292–293.

Chen BL, Hall DH, Chklovskii DB. (2006) Wiring optimization can relate neuronal structure and function. *PNAS* **103**(12): 4723–4728.

Chklovskii DB, Koulakov AA. (2004) Maps in the brain: What can we learn from them?' *Annu Rev Neuroscience* **27**: 369–392.

Chickering A. (2006) Strengthening spirituality and civic engagement in higher education. *J College Character* **8**(1) November.

Chong HQ, Tan AH, Ng GW. (2007) Integrated cognitive architecture — A survey. accepted for publication in *AI Rev J.*

Chopra S, Hadsell R, LeCun Y. (2005) Learning a similarity metric discriminatively, with application to face verification. *Proc IEEE Conf Computer Vision and Pattern Recognition.*

Chudler E. (2006) *Brain Facts and Figures.* Retrieved on July 39, 2006.

Chun MM. (1997) Types and tokens in visual processing: a double dissociation between the attentional blink and repetition blindness. *J Exp Psychol Hum Percept Perform.* **23**: 738.

Cohen MX, Ranganath C. (2007) Reinforcement learning signals predict future decisions. *J Neurosci* **27**(2): 371–378.

Collins FS. (2006) *The Language of God*. Free Press.

Cools R, Barker R, Sahakian B, Robbins T. (2001) *Cerebral* Cortex **11**: 1136.

Copeland J. Alan Turing: Code Breaker and AI Pioneer. *Talk in MIT on 30 Nov 2006.*

Covey SR. (1989) *The 7 Habits of Highly Effective People*. Simon and Schuster, New York.

Cox DD, Savoy RL. (2003) Functional magnetic resonance imaging (fMRI) 'brain reading': Detecting and classifying distributed patterns of fMRI activity in human visual cortex. *Neuroimage* **19**: 261–270.

Cox D, Meyers E, Sinha P. (2004) Contextually evoked object-specific responses in human visual cortex. *Science* **303**(5667): 115–117.

Coward, LA. (1999) A physiologically based theory of consciousness. In: Jordan S (ed) *Modeling Consciousness Across the Disciplines*, pp. 113–178, Maryland: UPA.

Csikszentmihalyi M, Rathunde K. (1990) The Psychology of wisdom; an evolutionary interpretation. In: Robert J Sternberg. *Wisdom: Its Nature, Origins and Development.*

Dale AM, Fischl B, Sereno MI. (1999) Cortical Surface-Based Analysis I. Segementation and Surface Reconstruction. *NeuroImage* **9**: 179–194.

Dalgleish T. (2004) The emotional brain. *Nat Rev, Neurosci* **5**: 582–589.

Daoyun J, Wilson MA. (2007) Coordinated memory replay in the visual cortex and hippocampus during sleep. *Nat Neurosci* **10**: 100–107.

Dayan P, Hinton G, Neal R, Zemel R. (1995) The Helmholtz machine. *Neural Comput* 7(8): 889–904.

Dayan P. (1999) Unsupervised learning. In: Wilson RA and Keil F (eds). *The MIT Encyclopedia of the Cognitive Sciences.*

Dayan P, Abbott LF. (2001) *Theoretical Neuroscience: Computational and Mathematical Modeling of Neural Systems.* Cambridge, Massachusetts: The MIT Press. ISBN 0-262-04199-5.

Dayan P, Balleine BW. (2002) Reward, motivation and reinforcement learning. *Neuron* **36**: 285–298.

Daw ND. (2003) *Reinforcement Learning Models of the Dopamine System and their Behavioral Implications.* Carnegie Mellon University.

Dean T. (2005) A computational model of the cerebral cortex. In: *Proceedings of AAAI-05,* pp. 938–943, Cambridge, MA. MIT Press.

DeJong KA, Samsonovich AV, Ascoli GA. (2006) An integrated self-aware cognitive architecture. *GMU Team BICA Phase 1, Phase I Preliminary Report.*

Dobbs D. (2006) Brain relies on stepwise activation to make difficult decisions. *Scientific American.com*: Oct 17, 2006.

Drubach D. (2000) *The Brain Explained.* Upper Saddle River, NJ: Prentice-Hall, Inc.

Dubash M. (2005) Moore's Law is dead, says Gordon Moore. *TechWorld,* 13 April 2005.

Dubuc (2006) Short term memory and long term memory. (n.d.). Retrieved October 23, 2006 from http://www.thebrain.mcgill.ca/flash/d/d_07/d_07_cr/d_07_cr_tra/d_07_cr_tra.html.

Duncan J. (1984) Selective attention and the organization of visual information. *J Exp Psch Gen* **113**: 501–517.

(Durbin and Willshaw, 1987)

Durbin R. Szeliski R, Yuille A. (1989) An Analysis of the elastic net approach to the traveling salesman problem. *Neural Comput* **1**: 348–358.

Durbin R, Mitchison G. (1990) A dimension reduction framework for understanding cortical maps. *Nature* **343**: 644–647.

Easter SS Jr, LS Ross, A Frankfurter. (1993) Initial track formation in the mouse brain. *J Neurosci* **13**(1): 285–299.

Edsinger AL. (2007) *Robot Manipulation in Human Environments.* PhD Thesis, MIT.

Edelman GM. (1989) *The Remembered Present: A Biological Theory of Consciousness.* Basic Books.

Eichenbaum H, NJ Cohen. (2001) *From Conditioning to Conscious Recollection: Memory systems of the brain.* Oxford University Press.

Eichenbaum H, Yonelinas AR, Ranganath C. (2007) The medial temporal lobe and recognition memory. *Annu Rev Neurosci* **30**: 123–152.

Eichenbaum H. (2007) *Comparative Cognition, Hippocampal Function, and Recollection*, **2**, pp. 47–66.

Ellias SA, Grossberg S. (1975) Pattern formation, contrast control, and oscillations in the short-term memory of shunting on-center off-surround networks. *Biol Cybernetics* **20**: 69–98. Available in PDF (EllGro1975BiolCyb.pdf) (1,923Kb).

Epstein R, Graham KS, Downing PE. (2003) Viewpoint-specific scene representations in human parahippocampal cortex. *Neuron* **37**(5): 865–876.

Eric R Kandel, James H Schwartz. (1985) *Principles of Neural Science, 2nd ed.* Elsevier.

Farley BJ. (2007) *Principles underlying the development and organisation of feature maps in the visual cortex*, PhD thesis, Feb 2007.

Fauconnier G, Turner M. (2002) *The Way We Think — Conceptual Blending and The Mind's Hidden Complexities.* Basic Book.

Fauconnier G, Turner M. (1998) Conceptual integration networks. **22**(2): 133–187.

Fazl A, S Grossberg, E Mingolla. (2005a). Invariant object learning and recognition using active eye movements and attentional control. *Proceedings of Ninth International Conference on Cognitive and Neural Systems*: 53.

Fazl A, S Grossberg, E Mingolla. (2005b) Invariant object learning and recognition using active eye movements and attentional control. *J Vision* **5**: 738–738.

Fazl A, S Grossberg, E Mingolla. (2006) View-invariant object category learning: How spatial and object attention are coordinated using surface-based attentional shrouds. *J Vision* **6**: 315.

Feigenbaum E, Feldman J (eds.) (1963) Computing machinery and intelligence, *Mind* **59**(1950): 433–460, reprinted in: Feigenbaum E, Feldman J (eds.), *Computers and Thought.* New York, McGraw-Hill.

Fei-Fei L, Fergus R, Perona P. (2004) Learning generative visual models from few training examples: An incremental Bayesian approach tested on 101 object categories. *Proc IEEE Conf Computer Vision and Pattern Recognition, Workshop Generative-Model Based Vision.*

Felleman D, DC Van Essen. (1991) Distributed hierarchical processing in the primata cerebral cortex. *Cerebral Cortex* **1**(1): 1–47.

Fergus R, Perona P, Zisserman A. (2003) Object class recognition by supervised scale-invariant learning. *Proc IEEE Conf Computer Vision and Pattern Recognition* **2**: 264–271.

Ferrer I, Fabregue I, Condom E. (1986) A Golgi study of the sixth layer of the cerebral cortex. *J Anatomy* **145**: 217–234.

Ferster D, Miller KD. (2000) Neural mechanisms of orientation selectivity in the visual cortex. *Annu Rev Neurosci* **23**: 441–471.

Fingelkurts An A, Fingelkurts Al A. (2004) Making complexity simpler: Multivariability and metastability in the brain. *Intl J Neurosci* **114**(7): 843–862. Url: http://www.bm-science.com/team/art30.pdf.

Fingelkurts An A, Fingelkurts Al A, Kähkönen SA. (2005) Functional connectivity in the brain — is it an elusive concept? *Neurosci Biobehav Rev* **28**(8): 827–836. Url: http://www.bm-science.com/team/art33.pdf.

Fischl B, Sereno MI, Dale AM. (1999) Cortical surface-based analysis II: Inflation, flattening, and a surface-based coordinate system. *NeuroImage* **9**: 195–207.

Fischl B, Salat DH, Busa E, Albert M, Dieterich M, Rosen B, Dale AM. (2002) Whole brain segmentation: Neurotechnique automated labeling of neuroanatomical structures in the human brain. Neuron **33**: 341–355.

Fortin NJ, Wright SP, Eichenbaum H. (2004) Recollection-like memory retrieval in rats is dependent on the hippocampus. *Nature* **431**: 188–190.

Fortin NJ, Agster KL, Eichenbaum H. Critical role of the hippocampus in memory for sequences of events. *Nat Neurosci* **May 5(5)**.

Frances K Skinner. (2006) Conductance-based models. In: *Scholarpedia*, pp. 5475.

Frank MJ, Seeberger LC, O'Reilly RC. (2004) By carrot or by stick: Cognitive reinforcement learning in Parkinsonism. *Sciencexpress* **Nov**.

Freeman WJ. Cortical aperiodic shutter enabling phase transitions at theta rates. *Proceedings International Joint Conference on Neural Networks, Orlando, FL, 12–17 August 2007.*

Freeman WJ. (1997) Three centuries of category errors in studies of the neural basis of consciousness and intentionality. *Neural Networks* **10**: 1175–1183.

Fukushima K. (1980) Neocognitron: A self-organizing neural network model for a mechanism of pattern recognition unaffected by shift in position. *Biol Cyberbetics* **36**: 193–201.

Fuster JM. (1997) *The Prefrontal Cortex*. Raven Press, New York.

Gallese V, Goldman A. (1998) Mirror neurons and the simulation theory of mindreading. *Trends Cog Sci* **2**: 493–501.

Gardner H. (1999) *Intelligence Reframed*. Basic Books.

Gardner H. (2003) Multiple intelligences after twenty years. *Paper presented at the American Educational Research Association, Chicago, Illinois, April 23, 2003.*

Gaiarsa JL, Caillard O, Ben-Ari Y. (2002) Long-term plasticity at GABAergic and glycinergic synapses: Mechanisms and functional significance. *Trends Neurosci* **25**(11): 564–570.

Gao E, Suga N. (1998) Experience-dependent corticofugal adjustment of midbrain frequency map in bat auditory system. *Pro Natl Acad Sci USA* **95**: 12663–12670.

Ge S, Yang C, Hsu K, Ming G, Song H. (2007) A critical period for enhanced synaptic plasticity in newly generated neurons of the adult brain. *Neuron* **54**: 559–566.

George D, Hawkins J. (2005) A hierarchical Bayesian model of invariant pattern recognition in the visual cortex. *Proceedings of the International Joint Conference on Neural Networks. IEEE.*

Georgeff MP, Ingrand FF. (1989) Decision-making in an embedded reasoning system. *Proceedings of the International Joint Conference of Artificial Intelligence*: 972–978.

Gloveli T, Dugladze T, Rotstein HG, Traub RD, Monyer H, Heinemann U, Whittington MA, Kopell NJ. (2005) Orthogonal arrangement of rhythm-generating microcircuits in the hippocampus. *PNAS* **102**(37): 13295–13300.

Gelernter D. (1994) *The Muse in the Machine*. The Free Press.

Gernsbacher MA. (1993) Less skilled readers have less efficient suppression mechanisms. *Psych Sci* **4**: 294–298.

Gerrig RJ, Zimbardo PG. (2005) *Psychology and Life, 17th ed.* Boston: Pearson and Allyn and Bacon.

Georgopoulos AP, Lurito JT, Petrides M, Schwartz AB, Massey JT. (1989) Mental rotation of the neuronal population vector. *Science* **243**: 234–236.

Georgopoulos AP, Pellizzer G, Poliakov AV, Schieber MH. (1999) Neural coding of finger and wrist movements. *J Comput Neurosci* **6**(3): 279–288.

Gibson J. (1979) *The Ecological Approach to Visual Perception.* Houghton Mifflin.

Gilbert C, Ito M, Kapadia M, Westheimer G. (2000) Interactions between attention, context and learning in primary visual cortex. *VISION RES 2000* **40**(10–12): 1217–1226.

Glimcher PW. (2004) *Decisions, Uncertainty, and the Brain (The Science of Neuroeconomics).* The MIT Press.

Glymour C. (2001) The *Mind's Arrows' Bayes Nets and Graphical Causal Models in Psychology. The MIT Press.*

Gopnik A, Nazzi T. (2003) Words, kinds and causal powers: A theory theory perspective on early naming and categorization. In: Rakison D, Oakes L (eds.), *Early categorization.* Oxford: Oxford University Press.

Gopnik A. (2003). The theory theory as an alternative to the innateness hypothesis. In: Antony L, Hornstein N (eds.), *Chomsky and His Critics.* Blackwells, Oxford.

Gopnik A, J Tenenbaum. (2007) Bayesian special section: Introduction — Bayesian networks, Bayesian learning and cognitive development. *Dev Sci* **10**(3): 281–287.

Gratch J, Marsella S, Mao W. (2006) Toward a Validated Model of Emotional Intelligence. *Twenty-First National Conference on Artificial Intelligence (AAAI06)* (Boston, MA).

Greenberg S, Kingsbury BED. (1997) The modulation spectrogram: In pursuit of an invariant representation of speech. *ICASSP-97, Munich* **3**: 1647–1650.

Gros Claudius. (2005) Self-sustained thought processes in a dense associative network. In: *Lecture notes in computer science, Vol. 3698, KI 2005: Advances in Artificial Intelligence.* Springer Berlin/ Heidelberg.

Grossberg S. (1968) Some physiological and biochemical consequences of psychological postulates. *Proc Natl Acad Sci* **60**: 758–765.

Grossberg S. (1971) On the dynamics of operant conditioning. *J Theor Biol* **33**: 225–255.

Grossberg S. (1972) A neural theory of punishment and avoidance, I: Qualitative theory. *Math Biosci* **15**: 39–67.

Grossberg S. (1972) Neural expectation: Cerebellar and retinal analogs of cells fired by learnable or unlearned pattern classes. *Kybernetik* **10**: 49–57.

Grossberg S. (1976) Adaptive pattern classification and universal recoding II: Feedback expectation, olfaction, and illusions. *Biol Cybernetics* **23**: 187–202.

Grossberg S. (1982) *Studies of Mind and Brain*. Dordrecht, The Netherlands: D Reidel Publishing Company.

Grossberg S. (1982) Associative and competitive principles of learning and development: The temporal unfolding and stability of STM and LTM patterns. In: Amari SI, Arbib M (eds.), *Competition and Cooperation in Neural Networks*, pp. 295–341. New York: Springer-Verlag.

Grossberg S. (1982) Processing of expected and unexpected events during conditioning and attention: A psychophysiological theory. *Psych Rev* **89**: 529–572.

Grossberg, S. (1982) Why do cells compete? Some examples from visual perception. *The UMAP Journal* **3**: 103–121. Available in PDF (Gro1982UMAPJ.pdf) (267Kb).

Grossberg S. (1984) Some normal and abnormal behavioral syndromes due to transmitter gating of opponent processes. *Biol Psych* **19**: 1075–1118.

Grossberg S. (1984) Some psychophysiological and pharmacological correlates of a developmental, cognitive and motivational theory. In: Karrer R, Cohen J, Tueting P (eds.), *Brain and Information: Event Related Potentials*, pp. 58–142. New York: New York Academy of Sciences.

Grossberg S. (1988) Nonlinear neural networks: Principles, mechanisms, and architectures. *Neural Networks* **1**: 17–61. Available in PDF (Gro1988NN.pdf)(11,497Kb).

Gerstner W, Kistler WM, (2002) *Spiking Neurons Models — Single neurons, Populations, Plasticity.* Cambridge University Press.

Grossber S, Levine DS. (1987) Neural dynamics of attentionally modulated Pavlovian conditioning: Block, inter-stimulus interval and secondary conditioning, *Applied Optics* **26**: 5015–5030.

Grossberg S, Merrill JWL. (1992) A neural network model of adaptively timed reinforcement learning and hippocampal dynamics. *Cog Brain Res* **1**: 3–38.

Grossberg S, Merrill JWL. (1996) The hippocampus and cerebellum in adaptively timed learning, recognition and movement. *J Cog Neurosci* **8**: 257–277.

Grossberg S. (1997) Neural dynamics of variable-rate speech categorization. *J Exp Psych: Human Perception and Performance* **23**: 481–503.

Grossberg S. (1998) Birth of a learning law. *INNS/ENNS/JNNS Newsletter* **21**: 1–4. Appearing with *Neural Networks* **11**(1).

Grossberg S. (1999) The link between brain learning, attention and consciousness. *Consciousness and Cognition* **8**: 1–44.

Grossberg S, Williamson JR. (1999) A self-organizing neural system for learning to recognize textured scenes. *Vis Res* **39**: 1385–1406.

Grossberg S. (2000) The complementary brain: Unifying brain dynamics and modularity. *Trends in Cognitive Sciences* **4**: 233–246.

Grossberg S. (2000) Linking mind to brain: The mathematics of biological intelligence. *Notices of the American Mathematical Society* **47**: 1361–1372.

Grossberg S. (2000) How hallucinations may arise from brain mechanisms of learning, attention and volition. *J Intl Neuropsych Society* **6**: 579–588.

Grossberg S, Mingolla E, Viswanathan L. (2001) Neural dynamics of motion integration and segmentation within and across apertures. *Vis Res* **41**: 2351–2553.

Grossberg S. (2003) Linking visual cortical development to visual perception. In: Hopkins B, Johnson SP (eds.), *Advances in Infancy Research: Neurobiology of Infant Vision.* Westport. CN: Praeger Publishers, in press.

Grossberg S. (2003) How does the cerebral cortex work? Development, learning, attention, and 3D vision by laminar circuits of visual cortex. *Behav Cog Neurosci Rev*, in press.

Grossberg S. (2003) Resonant neural dynamics of speech perception. *J Phonetics*, in press.

Grossberg S, Govindarajan KK, Wyse LL, Cohen MA. (2004) ART-STREAM: A neural network model of auditory scene analysis and source segregation. *Neural Networks* **17**: 511–536.

Grossberg S. (2005) Linking attention to learning, expectation, competition, and consciousness. In: Itti L, Rees G, Tsotsos J (eds.), *Neurobiology of Attention*, pp. 652–662. San Diego: Elsevier.

Grossberg S, Yazdanbakhsh A. (2005) Laminar cortical dynamics of 3D surface perception: Stratification, transparency, and neon color spreading. *Vis Res* **45**: 1725–1743.

Grossberg S, Pearson L. (2006) Laminar cortical dynamics of cognitive and motor working memory, sequence learning and performance: Towards a unified theory of how the cerebral cortex works. *CAS/CNS Tech Report-2006–2002*. Submitted for publication.

Grossberg S, Seidman D. (2006) Neural dynamics of autistic behaviours: Cognitive, emotional and timing substrates. *Psych Rev* **113**: 483–525.

Grossberg S. (2006) Towards a unified theory of neocortex: Laminar cortical circuits for vision and cognition. *Technical Report CAS/CNS-2006–2008.*

Grossberg S, Kuhlmann L, Mingolla E. (2007) A neural model of 3D shape-form-texture: Multiple-scale filtering, boundary grouping, and surface filling-in. *Vis Res, ScienceDirect* **47**: 634–672.

Hawkins J. (2004) *On Intelligence*. New York: Times Books.

Hawkins J. (2007) Learn like a human. *IEEE Spectrum*, April 2007.

Haxby JV, Hoffman EA, Gobbini MI. (2000) The distributed human neural system for face perception. *Trends Cog Sci* **4**(6): 223–233.

He S, Cavanagh P, Intriligator J. (1996) Attentional resolution and the locus of visual awareness. *Nature* 383, 334.

Heeger DJ. (1992) Normalization of cell responses in cat striate cortex. *Vis Neurosci* **9**: 198.

Heekeren HR, Marrett S, Bandettini PA, Ungerleider LG. (2004) A general mechanism for perceptual decision-making in the human brain. *Nature* **431**: 859–862. (PubMed)

Heekeren HR, Marrett S, Ruff DA, Bandettini PA, Ungerleider LG. (2006) Involvement of human left dorsolateral prefrontal cortex in perceptual decision making is independent of response modality. *Proc Natl Acad Sci USA* **103(26)**: 10023–10028.

Helper AN. (2003) The other side of the story. *Psychiatric Times*, June 2003.

Hikosaka O, Takikawa Y, Kawagoe R. (2000) Role of the Basal Ganglia in the Control of Purposive Saccadic Eye Movements. *Physiological Reviews* **80**(3).

Hirsch JA, Alonso JM, Reid RC, Martinez LM. (1998) Synaptic integration in striate cortical simple cells. *J Neurosci* **18**(22): 9517–9528.

Hodgkin AL, Huxley AF. (1952) A quantitative description of membrane current and its application to conduction and excitation in nerve. *J Physiol* **117**: 500–544.

Holden Constance. (2007) Fresh teeth from the lab. *ScienceNOW Daily News, February 2007.*

Horowitz TS, Choi WY, Horvitz JC, Cote LJ, Mangel JA. (2006) Visual search deficits in Parkinson's disease are attenuated by bottom-up target salience and top-down information. *Neuropsychologia* **44**(10): 1962–1977.

Horowitz TS, Birnkrant RS, Fencsik DE, Tran L, Wolfe JM. (2006) How do we track invisible objects? *Psychon Bull Rev* **Jun 13**(2): 516–523.

Horst Steven. (2005) The computational theory of mind. *Stanford Encyclopedia of Philosophy.*

Horwitz B. (2003) The elusive concept of brain connectivity. *NeuroImage* **19**: 466–470.

Hoshi Y. (2003) Functional near-infrared optical imaging: Utility and limitations in human brain mapping. *Psychophysiology* **40**: 511–520.

Hostler Levere C. (1982) An SL (2,C)-invariant representation of the Dirac equation. *J Math Phys* **23**: 1179.

Hubel DH, Wiesel TN. (1962) Receptive fields, binocular interaction and functional architecture in the cat's visual cortex. *J Physio* **160**: 106–154.

Hubel DH, Wiesel TN. (1974) *J Comp Neurol* **158**: 295–305.

Hubel DH, Wiesel TN. (1977) *Proc R Soc Londaon Ser B* **198**: 1–59.

Hodgkin AL, Huxley AF. (1952) A quantitative description of membrane current and its application to conduction and excitation in nerve. *J Physiol (London)* **117**: 500–544.

Hurdal MK, Kurtz KW, Banks DC. (2001) Case study: Interacting with cortical flat maps of the human brain. In: *Proceedings Visualization 2001,* pp. 469–472 & 591. IEEE, Piscataway, NJ.

Iacoboni M, Woods RP, Brass M, Bekkering H, Mazziotta JC, Rizzolatti G. (1999) Cortical mechanisms of human imitation. *Science* **286**: 5449.

Isard M. (2003) PAMPAS: Real-valued graphical models for computer vision. In: *Proceedings of the 2003 IEEE Computer Society Conference on Computer Vision and Pattern Recognition* **I**: 613–620.

Isen AM. (2000) Positive affect and decision making. In: Lewis M, Haviland J (eds), *Handbook of Emotions.* Guilford, New York.

Izhikevich EM. (2003) Simple model of spiking neurons. *IEEE Transactions of neural networks* **14**(6): 1569–1573.

James William. (1890) *The Principles of Psychology.* New York: Henry Holt.

Jezzard P, Clare S, (1999) Sources of distortion in functional MRI data. *Human Brain Mapping* **8**: 80–85.

Johansson C, Lansner A. (2007) Towards cortex sized artificial neural systems. *Neural Network* **20**(1): 48–61.

Kanwisher N, McDermott J, Chun MM. (1997) The fusiform face area: a module in human extrastriate cortex specialized for face perception. *J Neurosci* **17**(11): 4302–11.

Kandel ER, Schwartz JH, Jessel TM. (2000) Central visual pathways. In: *Principles of Neural Science,* Ch 27. McGraw-Hill, New York.

Kane MJ, Hambrick DZ, Tuholski SW, Wilhelm O, Payne TW, Engle RE. (2004) The henerality of working memory capacity: A latent variable approach to verbal and visuospatial memory span and reasoning. *J Exp Psych: General* **133**: 189–217.

Karmarkar UR, Buonomano DV. (2003) Temporal specificity of perceptual learning in an auditory discrimination task. *Learning and Memory* **10**(2): 141–147.

Kay LM. (2003) A challenge to chaotic itinerancy from brain dynamics. *Chaos* **13**(3): 1057–1066.

Khosla D. (2006) Biologically-inspired Cognitive Architecture for Integrated learning, Action and Perception (BICA-LEAP). *DARPA BICA Project.*

Kim DS, Ronen I, Olman C, Kim SG, Ugurbil K, Toth LJ. (2004) Spatial relationship between neuronal activity and BOLD functional MRI. *Neuroimage* **21**(3): 876–885.

Kish S, Shannak K, Hornykiewicz O. (1988) *New England J Med*: 318, 876.

Knowlton BJ, Margels JA, Squire LR. (1996) *Science* 273, 1399.

Kolb B, Whishaw IQ. (1996) *Fundamentals of Human Neuropsychology.* Lethbridge: Worth Publishers.

Kolb B, Whishaw IQ. (2006) *An Introduction to Brain and Behavior, 2nd ed.* New York: Worth Publishers.

Kornack DR, Rakic P. (1995) Radial and horizontal deployment of clonally related cells in the primate neocortex: Relationship to distinct mitotic lineages. *Neuron* **15**(2): 311–321.

Kringelbach ML. (2005) The orbitofrontal cortex: Linking reward to hedonic experience. *Nat Rev Neurosci* **6**: 691–702.

Krupa DJ, Ghazanfar AA, Nicolelis MA. (1999) Immediate thalamic sensory plasticity depends on corticothalamic feedback. *Proc Natl Acad Sci USA* **96**: 8200–8205.

Kwong KK, Belliveau JW, Chesler DA, Goldberg IE, Weisskoff RM, Poncelet BP, Kennedy DN, Hoppel BE, Cohen MS, Turner R, Cheng H, TJ Brady, Rosen BR. (2003) *Dynamic Magnetic Resonance Imaging of Human Brain Activity During Primary Sensory Stimulation,* **89**, pp. 5675–5679.

Lai CSL, Fisher SE, Hurst JA, Vargha-Khadem F, Monaco AP. (2001) A forkhead-domain gene is mutated in a severe speech and language disorder. *Nature* **413**.

Laird JE, Newell A, Rosenbloom PS, (1987) An architecture for general intelligence. *Artificial Intelligence* **33**: 1–64.

Laird JE, Rosenbloom PS, Newell A. (1986) Chunking in Soar: The anatomy of a general learning mechanism. *Machine Learning* **1**: 11–46.

Laird JE, Rosenbloom PS, Newell A. (1986) *Universal Subgoaling and Chunking: The Automatic Generation and Learning of Goal Hierarchies*. Kluwer. Boston, MA.

Lang S. (1996) Psychologist find dopamine linked to a personality traits and happiness. *Cornell Research*.

Langley P, Arai S, Shapiro D. (2004) Model-based learning with hierarchical relational skills. *Proceedings of the ICML-2004 Workshop on Relational Reinforcement Learning*.

Langley P, Choi D. (2006) A unified cognitive architecture for physical agents. *Proceedings of the Twenty-First National Conference on Artificial Intelligence*: 1469–1474.

Lansner A, Holst A. (1996) A higher order Bayesian neural network with spiking units. *Intl J Neural Sys* 7(2): 115–128.

Laurent G, Stopfer M, Friedrich R, Rabinovich MI, Volkovskii A, Abarbanel HDI, (2001) Odor encoding as an active, dynamical process: Experiments, computation and theory. *Annu Rev Neurosci* **24**: 263–297.

Lavenex P, Lavenex PB, Amaral DG. (2006) Spatial relational learning persists following neonatal hippocampal lesions in macaque monkeys. *Nat Neurosci* **10**: 234–239.

Lee TS, David M. (2003) Hierarchical Bayesian Inference in the visual cortex. *J Optical Society of America* **2**(7): 1434–1448.

Lee Gutkind. (2006) *Almost Human: Making Robots Think*. Norton & Company Publisher.

LeBihan D. (2007) The 'wet mind': Water and functional neuroimaging. *Physics Med Biol* **52**: R57–R90.

LeCun Y, Bottou L, Bengio Y, Haffner P. (1998) Gradient-based learning applied to document recognition. *Proc IEEE* **86**(11): 2278–2324.

LeCun Y, Huang FJ, Bottou L. (2004) Learning methods for generic object recognition with invariance to pose and lighting. *Proc IEEE Conf Computer Vision and Pattern Recognition*.

LeDoux JE. (1994) Emotion, memory and the brain. *Scientific American* **June**: 50–57.

Lee SH, Blake R, Heeger DJ. (2005) Traveling waves of activity in primary visual cortex during binocular rivalry. *Nat Neurosci* **8**(1).

Lehman JF, Laird JE, Rosenbloom PS. (2006) *A gentle introduction to soar, an architecture for human cognition: 2006 update.*

Lerner JS, Keltner D. (2000) Beyond valence: Toward a model of emotion-specific influences on judgement and choice. *Cognition and Emotion* **14**: 473–493.

Li W, Piech V, Gilbert CD. (2004) Perceptual learning and top-down influences in primary visual cortex *Nat Neurosci* 7(6): 651–657.

Li Y, Ma Z, Lu W, Li Y. (2006) Automatic removal of the eye blink artifact from EEG using an ICA-based template matching approach. *Physiol Measurement* **27**: 425–436.

Liegeois F, Baldeweg T, Connelly A, Gadian DG, Mishkin M, Vargha-Khadem F. (2003) Language fMRI abnormalities associated with FOXP2 gene mutation. *Nat Neurosci* **6**(11).

Lori NF, Akbudak E, Shimony JS, Cull TS, Snyder AZ, Guillory RK, Conturo TE. (2002) Diffusion tensor fiber tracking of human brain connectivity: Acquisition methods, reliability analysis and biological results. *NMR Biomed* **15**: 494–515.

Louie K, Wilson MA. (2001) Temporally structured REM sleep replay of awake hippocampal ensemble activity. *Neuron* **29**: 145–156.

Loftus EF. (2003) Make-believe memories. *American Psychologist* **Nov**: 867–872.

Loftus EF. (2003) Our changeable memories: Legal and practical implications. *Nat Rev: Neurosci* **4**: 231–234.

Loftus EF. (2002) Memory faults and fixes. *Issues Sci Tech Natl Acad Sci* **18**.

Logothetis NK, Wandell BA. (2004) Interpreting the BOLD signal. *Annu Rev Phyciol* **66**: 735–769.

Lustig C, May CP, Hasher L. (2001) Working memory span and the role of proactive interference. *J Exp Psych: General* **130**: 199–207.

Mack A, Rock I. (1998) *Inattentional Blindness.* MIT Press, Cambridge, MA.

Maguire EA, Spiers HJ, Good CD, Hartley T, Frackowiak RSJ, Burgess N. (2003) Navigation expertise and the human hippocampus: A structural brain imaging analysis. *Hippocampus* **13**(2): 250–259.

Malenka RC, Nicoll RA. (1999) Long-Term Potentiation — A Decade of Progress? *Rev Sci* **285**(5435): 1870–1874.

Majewska A, Sur M. (2006) Plasticity and Specificity of Cortical Processing Networks. *Trends Neurosci* **29**(6).

Mandavilli A. (2006) Visual neuroscience: Look and Learn. *Nature* **441**(7091): 27 1–2.

Mandler G. (1985) Cognitive Psychology: An Essay in Cognitive Science. Hillsdale, NJ:Erlbaum.

Markram H. (2006) The Blue Brain Project. *Nat Rev Neurosci* **7**: 153–160.

McCarthy C, Mejia OL, Liu HT, Durham AC. (1998) Cognitive appraisal theory: A psychoeducational model for connecting thoughts and feelings. *Multicultural Group Work. Paper presented at Annual Convention of the American Psychological Association.*

McClelland JL, Thompson RM. (2007) Using domain-general principles to explain children's causal reasoning abilities. *Dev Sci* **10**(3): 333–356.

McClelland JL, O'Reilly RC, McNaughton BL. (1995) Why there are complenentary learning systems in the hippocampus and neocortex: Insights from the successes and failures of connectionist models of learning and memory. *Psych Rev.*

Merkle RC. (1989) Energy limits to the computational power of the human brain. *Foresight Update (NanoTech Institute)* **1**.

Merkle RC. (1988) How many bytes in human memory. *Foresight Update* **4**.

Merker B. (2006) Consciousness without a cerebral cortex: A challenge for neuroscience and medicine. *Behav Brain Sci* (in press).

Merzenich MM, Jenkins WM, Johnston P, Schreiner C, Miller SL, Tallal P, (1996) Temporal processing deficits of language-learning impaired children ameliorated by training. Science **271**: 77–81.

Metz C, Bsales J. (2007) Five Ideas that will Reinvent Modern Computing. *PC Magazine* **June**.

Miller GA. (1956) The magical number seven, plus or minus two: Some limits on our capacity for processing information. *Psych Rev* **63**: 81–97.

Minsky Marvin. (1986) *The Society of Mind.* Simon and Schuster.

Minsky Marvin. (2006) *The Emotion Machine.* Simon and Schuster.

Moore JK, Guan YL, Shi SR. (1996) Axogenesis in the human fetal auditory system, demonstrated by neurofilament immunohistochemistry. *J Anatomy Embryol* **195**(1).

Mountcastle VB. (1957) Modality and topographic organization of single neurons of cat's somatic sensory cortex. *J Neurophysiol* **20**: 408–434.

Mountcastle VB. (1997) The columnar organization of the neocortex. *Brain, Review Article* **120**: 701–722.

Moravec H. (1998) *Robot, Being: Mere Machine to Transcendent Mind.* Oxford University Press.

Moravec H. (1998) When will computer hardware match the human brain? *J Evol Tech* **1**.

Mumford D. (1992) On computational architecture of the neocortex. *Biol Cybernetics* **66**(3): 241–251.

Myers RE, Sperry RW. (1958) Interhemispheric communication through the corpus callosum. *Archives Neuro Psych* **80**: 298–303.

Nelson ME. (2004) Electrophysiological models. In: Koslow S, Subramaniam S (eds), *Databasing the Brain: From Data to Knowledge.* Wiley, New York.

Newell A. (1990) *Unified Theories of Cognition.* Harvard University Press, Cambridge, MA.

Newton JR, M Sur. (2007) Developmental studies on rewiring the brain: What they tell us about brain evolution. In: Kaas JH (ed), *Evolution of Nervous Systems.* **3,** pp. 103–112. Academic Press, Oxford.

Ng GW. (1997) *Application of Neural Network to Adaptive Control of Nonlinear Systems.* Research Studies Press.

Ng GW. (2003) *Intelligent Systems — Fusion, Tracking and Control.* Research Studies Press.

Novojilova AP, Babmindra VP. (2004) Intracortical horizontal connections of neurons in cat and monkey motor cortex. *J Neurosci* **3**(5/6): 389–396.

O'Regan JK, Rensink RA, Clark JJ. (1999) Change-blindness as a result of 'mudsplashes'. *Nature* **398**: 34.

O'Reilly RC. (1996) Biologically plausible error-driven learning using local activation differences: The generalised recirculation algorithm. *Neural Comput* **8**: 895–938.

O'Reilly RC. (1998) Six principles for biologically-based computational models of cortical cognition. *Trends Cog Sci* **2**: 455–462.

O'Reilly RC, Frank MJ. (2006) Making working memory work: A computational model of learning in the frontal cortex and basal ganglia. *Neural Comput* **18**: 283–328.

Ormrod JE. (1999) *Human Learning, 3rd Ed.* Prentice-Hall, Upper Saddle River, NJ.

O'Keefe J, Recce ML. (1993) *Hippocampus* 3, pp. 317–330.

Olive JP. (1996) The talking computer — Text to speech synthesis. In: David Stork (ed), *HAL's Legacy.*

Ohki K, Chung S, Chng YH, Kara P, Reid RC. (2005) Functional imaging with cellular resolution reveals precise micro-architecture in visual cortex. *Nature* **433**: 597–603.

Ohki K, Chung S, Kara P, Hubener M, Bonhoeffer T, Reid RC. (2006) Highly ordered arrangement of single neurons in orientation pinwheels. *Nature* **442**: 925–928.

Orre *et al.* (2000) Bayesian neural networks with confidence estimations applied to data mining. *Comput Stat Data Analysis* **34**(4): 473–493.

Ortony A, Clore GL, Collins A. (1988) *The Cognitive Structure of Emotions.* Cambridge University Press, Cambridge.

Owen-Smith P. (2004) What is cognitive-affective learning? *J Cog Affective Learning* **1**(1).

Ozarslan E, Vemuri B, Mareci T. (2004) Fiber orientation mapping using generalized diffusion tensor imaging. *Biomedical Imaging: Macro to Nano (International Symposium)* **1**: 1036–1039.

Papert S. (2005) You can't think about thinking without thinking about thinking about something. *Contemporary Issues Tech Teacher Educ* 5(3/4): 366–367.

Penrose R. (1991) *The Emperor's New Mind.* Oxford University Press, 1st Print 1989, Penguin Books, Reprinted Edition 1991.

Perrett D, Oram M. (1993) Neurophysiology of shape processing. *Img Vis Comput* **11**: 317–333.

Picard R. (1996) Does HAL cry digital tears? Emotions and computers. In: *'HAL's Legacy — 2001's Computer as Dream and Reality.* MIT Press.

Picard RW, S Papert, Bender W, Blumberg B, Breazeal C, Cavallo D, Machover T, Resnick M, Roy D, Strohecker C. (2004) Affective learning — a manifesto. *BT Tech J* **22**(4).

Pilly P, Grossberg S. (2005) Brain without Bayes: Temporal dynamics of decision making in the laminar circuits of visual cortex. *Society of Neuroscience Abstracts* **591**(1).

Plutchik R. (1962) *The Emotions: Fact, Theories, and a New Model.* New York Random House.

Plutchik R. (2002) *Emotions and Life: Perspectives for Psychology, Biology, and Evolution.* Book Forum Mind/Brain. Washington, DC, American Psychological Association.

Polimeni J, Schwartz E. (2002) Neural representation of sensory data. Commentary on 'Mental imagery: In search of a theory' by Zenon W Pylyshyn. *Behav Brain Sci* **25(2)**: 207–208.

Polimeni JR, Granquist-Fraser D, Wood RJ, Schwartz EL. (2005) Physical limits to spatial resolution of optical recording: Clarifying the spatial structure of cortical hypercolumns. *PNAS* **102**(11).

Polimeni JR, M Balasubramanian, Schwartz EL. (2006) Multi-area visuotopic map complexes in macaque striate and extra-striate cortex. *Vis Res* **46**: 3336–3359.

Pray L. (2003) fMRI The Perfect Imperfect Instrument. *The Scientist* **17**(21): 40.

Quirk GJ. (2002) Memory for extinction of conditioned fear is long-lasting and persists following spontaneous recovery. *Learning and Memory* **9**(6): 402–407. Cold Spring Harbor Laboratory Press.

Quiroga QR, Reddy L, Kreiman G, Koch C, Fried I. (2005) Invariant visual representation by single neurons in the human brain. *Nature* **435**: 1102–1107.

Rabinowicz T, Petetot JM, Gartside PS, Sheyn D, deCaurten-Myers GM. (2002) Structure of the cerebral cortex in men and women. *J Neuropath Exp Neurol* **61**(1): 46–57.

Rao RPN, Ballard DH. (1996) Dynamic model of visual recognition predicts neural response properties in the visual cortex. *Neural Comput* **9**: 721–763.

Rao AS, Georgeff MP. (1991) Modeling rational agents within a BDI-architecture. In: *Proceedings, Second International Conference on*

Principles of Knowledge Representation and Reasoning, pp. 473–484, Morgan Kaufmann, San Mateo, CA.

Raymond K. (1999) *The Age of Spiritual Machines.* Penguin Books Ltd.

Riesenhuber M, Poggio T. (1999) Hierarchical models of object recognition in cortex. *Nat Neurosci* **2**(11): 1019–1025.

Reilly RC. (1998) Six principles for biologically based computational models of cortical cognition. *Rev Trends Cog Sci* **2**(11).

Reilly RC. (2006) Biological based computational models of high-level cognition. *Science* **314**: 91–94.

Rendle-Short J. (1981) Man: The image of God. *Creation* **4**(1): 21–29.

Rojer AS, Schwartz EL. (1990) Design considerations for a space-variant visual sensor with complex-logarithmic geometry. *10th International Conference on Pattern Recognition* **2**: 278–285.

Rojer AS, Schwartz EL. (1990) Cat and monkey cortical columnar patterns modeled by bandpass-filtered 2D white noise. *Biol Cybernetics* **62**: 381–391.

Rokni U, Richardson AG, Bizzi E, Seung S. (2007) Motor learning with unstable neural representations. *Neuron* **54**: 653–666.

Roseman IJ, Antoniou AA, Jose PE. (1996) Appraisal determinants of Emotions: Constructing a more accurate and comprehensive theory. *Cognition and Emotion* **10**(3): 241–277.

Rumelhart DE, McClelland JL (eds.) (1986a) *Parallel Distributed Processing* **1,** MIT Press.

Rumelhart DE, McClelland JL (eds.) (1986b) *Parallel Distributed Processing* **2,** MIT Press.

Rumelhart DE, McClelland JL (eds.) (1987a) *Parallel Distributed Processing* **1,** MIT Press.

Rumelhart DE, McClelland JL (eds) (1987b) *Parallel Distributed Processing* **2**, MIT Press.

Russel JA. (1980) A circumplex model of affect. *J Personality Social Psych* **39**: 1161–1178.

Russell JA, Fehr B. (1987) Relativity in the perception of emotion in facial expressions. *J Exp Psych*, **116**(3): 223–237.

Russell JA, Barrett LF. (1999) Core affect, prototypical emotional episodes, and other things called emotion: Dissecting the elephant. *J Personality Social Psych* **76**(5): 805–819.

Russel J. (2003) Core affect and the psychological construction of emotion. *Psych Rev* **110**(1): 145–172.

Salvoey P, Mayer JD. (1990) Emotional intelligence. *Imagination, Cognition and Personality* **9**(3): 185–211.

Sanders R (2006). Slow brain waves play key role in coordinating complex activity. *UC Berkeley News*, Press Release, 14 September 2006.

Sarfati J. (1998) Answers in Genesis. *Creation Ex Nihilo Technical Journal* **12**(1): 20–22.

Saxe R, Wexler A. (2005) Making sense of another mind: The role of the right temporo-parietal junction. *Neuropsychologia.*

Saxe MD, Malleret G, Vronskaya S, Mendez I, Garcia AD, Sofroniew MV, Kandel ER, Hen R. (2007) Paradoxical influence of hippocampal neurogenesis on working memory. *PNAS* **104**: 4642–4646.

Scaruffi P. (1999) Review: Jonathan Winson. (1985) *Brain and Psyche.* Anchor Press.

Schaffhausen J. (1997) *The Day His World Stood Still.* Brainconnect.com, Scientific Learning Corporation.

Schank R. (1992) Story-based memory. (1992) Morelli R, Brown WM, Anselmi D, Haberlandt K, Lloyd D. (eds.), *Minds, Brains and Computers*, ch. 5. Ablex Publishing Corporation.

Schroeder G. (2000) The hidden face of God — How science reveals the ultimate truth. *Aishaudio.com.*

Schultz W, Dayan P, Montague PR. (1997) A neural substrate of prediction and reward. *Science* **275**: 1593–1599.

Schwart EL. (1977) Spatial mapping in the primate sensory projection: Analytic structure and relevance to perception. *Biol Cybernetics* **25**: 181–194.

Schwartz EL, Christman DR, Wolf AP. (1984) Human primary visual cortex topography imaged via positron tomography. *Brain Res* **294**(2): 225–230.

Schwartz EL, Merker B, Wolfson E, Shaw A. (1988) Applications of computer graphics and image processing to 2D and 3D modeling of the functional architecture of visual cortex. In: *Comput Neurosci IEEE Computer Graphics and Applications* **13**.

Schwartz EL, Shaw A, Wolfson E. (1989) A numerical solution to the generalized mapmaker's problem: flattening nonconvex polyhedral

surfaces. *IEEE Transcations on Pattern Analysis and Machine Intelligence* **11**(9).
Searle John R. (1980) Minds, brains, and programs. *Behavioral and Brain Sciences* **3(3)**: 417–457.
Searle JR. (1984) *Minds, Brains and Science.* Harvard University Press, hardcover: ISBN 0-67457631-4
Searle JR. (1990) Is the brain a digital computer? *Proceedings and Addresses of the American Philosophical Association* **64**: 21–37.
Searle JR. (1991) *The Rediscovery of the Mind.* MIT Press.
Serre T, Wolf L, Bileshi S, Riesenhuber M, Poggio T. (2007) Robust object recognition with cortex-like mechanisms. *IEEE Transactions on Pattern Analysis and Machine Intelligence* **29**(3).
Serre T, Oliva A, Poggio T. (2006) A feedforward theory of visual cortex accounts for human performance in rapid categorization. *CBCL Paper # MMVI-02*, Massachusetts Institute of Technology, Cambridge, MA.
Serre T, Oliva A, Poggio T. (2007) A feedforward architecture accounts for rapid categorization. *PNAS* **104**(15): 6424–6429.
Searle JR. (2005) Consciousness — what we still do not know? *The New York Review of Book* **52**(11).
Seitz A, Watanabe T. (2005) A unified model for perceptual learning. *Trends Cog Sci* **9**(7).
Seung HS. (1996) How the brain keeps the eyes still. *Proceeding of National Academic Science (Neurobiology)* **93**: 13339–13344.
Seung HS. (2003) Learning in spiking neural networks by reinforcement of stochastic synaptic transmission. *Neuron* **40**(6): 1063–1073.
Shallit JO. (2000) Can Computer Think. *Talk for the Kitchener-Waterloo-Cambridge-Guelph Humanists, November 20, 2000.*
Shannon RV, Zeng FG, Kamath V, Wygonski J, Ekelid M. (1995) Speech recognition with primarily temporal cues. *Science* **270**: 303–304.
Shepard RN. (1967) Recognition memory for words, sentences, and pictures. *J Verbal Learning and Verbal Behavior* **6**: 156–163.
Shrobe H, Winston P, Tennenbaum J, Shaftoe P, Massaquoi S, Williams B, Eslick I, Rao S, Coen M, Bobrow R. (2006) CHIP: A cognitive architecture for comprehensive human intelligence and performance. *DARPA BAA, BICA Program Proposal.*

Shoker L, Sanei S, Latif MA. (2004) Removal of eye blinking artifacts from EEG incorporating a new constrained BSS algorithm. *Engineering in Medicine and Biology Society*, 26th Annual International Conference of the IEEE **1, 2**: 909–912.

Siapas AG, Wilson MA. (1998) Coordinated interactions between hippocampal ripples and cortical spindles during slow-wave sleep. *Neuron* **21**: 1123–1128.

Singer TD. (2003) Neural population codes. *Curr Opin Neurobiol* **13(2)**: 238–249. Review. PMID: 12744980 PubMed — indexed for MEDLINE.

Singh P. (2005) EM-ONE: An architecture for reflective commonsense thinking. *PhD Thesis.*

Sit YF, Miikkulainen R. (2006) Self-organization of hierarchical visual maps with feedback connections. *Neurocomputing* **69**: 1309–1312.

Skarda CA, Freeman WJ. (1987) How brain makes chaos in order to make sense of the world. *Behav Brain Sci* **10**: 161–195.

Smith J. (1956) Sensations and brain processes. *Philosophical Rev.*

Smith CUM. (1970) *The Brain: Towards an Understanding.* New York: GP Putnam's Sons.

Smith AT, Singh KD, William AL, Greenlee MW. (2001) Estimating receptive field size from fMRI data in human striate and extrastriate visual cortex. *Cerebral Cortex* **11**: 1182–1190.

Snowden JS, Bathgate D, Varma A, Blackshaw A, Gibbons ZC, Neary D. (2001) Distinct behavioral profiles in frontotemporal dementia and semantic dementia. *J Neurol Neurosurg Psychiatry* **70**: 323–332.

Sowell ER, Thompson PM, Leonard CM, Welcome SE, Kan E, Toga AW. (2004) Longitudinal mapping of cortical thickness and brain growth in normal children. *J Neurosci* **24**(38): 8223–8231.

Sperry RW, Gazzaniga MS, Bogen JB. (1969) Interhemispheric relationships: The neocortical commissures: syndromes of hemisphere disconnection. In: PJ Vinken & GW Bruyn (eds), *Handbook of Clinical Neurology* **4**, pp. 273–290. Amsterdam: North-Holland.

Spetner Lee. (1997) *Not by Chance.* The Judaica Press Inc, Brooklyn, New York.

Spurgeon Chris. (2006) Souls of a new machine — How can you make a computer think like a person? *Boston Globe*: November 26, 2006.

Stein L, Belluzzi JD. (1989) Cellular investigations of behavioral reinforcement. *Neurosci Biobehav Rev* **13**: 69–80.

Stein L. (1994) *In-vitro* reinforcement of hippocampal bursting: Possible cellular and molecular mechanism of drug reward. *Regulatory Peptides* **54**: 285–286.

Stein L. (1997) Biological substrates of operant conditioning and the operant-respondent distinction. *J Exp Analysis Behav* **67**: 246–253.

Sternberg RJ. (2003) *Wisdom, Intelligence and Creativity Synthesized.* Cambridge University Press.

Stork DG. (1996) HAL's legacy — 2001's computer as dream and reality. MIT Press.

Sudderth E, Ihler A, Freewan W, Willsky A. (2003) Nonparametric belief propagation. In: *Proceedings of the IEEE Computer Society Conference on Computer Vision and Pattern Recognition,* **1**, pp. 605–612.

Sun R, Peterson T. (1996) Hybrid learning incorporating neural and symbolic processes. In: (1998) *IEEE International Conference on Fuzzy Systems*, pp. 727–732.

Sun R, Peterson T. (1996) Learning in reactive sequential decision tasks: The CLARION model. In: *IEEE International Conference on Neural Networks*, pp. 1073–1078.

Sun R, Peterson T. (1998) Hybrid learning incorporating neural and symbolic processes. In: *IEEE International Conference on Fuzzy Systems*, pp. 727–732.

Sun R, Merrill E, Peterson T. (2001) From implicit skills to explicit knowledge: A bottom-up model of skill learning. *Cog Sci* **25**(2): 203–244.

Sun R, Zhang X. (2006) Accounting for a variety of reasoning data within a cognitive architecture. *J Exp Theor AI* **18**(2): 169–191.

Suplee C. (2000) Key brain growth goes on into teens. *Washington Post*: 9 March 2000.

Sur M, Palars SL, Roe AW. (1990) Cross-modal plasticity in cortical development: Differentiation and specification of sensory neocortex. *Trends Neurosci* **13**: 227–233.

Sur M, Leamey CA. (2001) Development and plasticity of cortical areas and networks. *Nat Rev (Neurosci)* **2**: 251–255.

Sur M. (2001) Cortical development: Transplantation and rewiring studies. *International Encyclopedia of the Social and Behavioral Sciences* **4**: 2837–2842.

Sutton RS, Barto AG. (1998) *Reinforcement Learning*. MIT, Cambridge MA.

Staddon JER, Zhang Y. (1991) On the assignment-of-credit problem in operant learning. In: Commons ML, Grossberg S, Staddon JER (eds.), *Neural Network Models of Conditioning and Action: A Volume in the Quantitative Analyses of Behavior Series,* ch. 11, pp. 279–293. Lawrence Erlbaum, Hillsdale, NJ.

Stein L. (1994) In-vitro reinforcement of hippocampal bursting: Possible cellular and molecular mechanism of drug reward. *Reg Peptides* **54**: 285–286.

Stein L, Xue BG, Belluzzi JD. (1994) In vitro reinforcement of hippocampal bursting: A search for Skinner's atoms of behavior. *J Exp Analysis Behav* **61**: 155–168.

Stein L. (1997) Biological substrate of operant conditioning and the operant-respondent distinction. *J Exp Analysis Behav* **67**: 246–253.

Stuphorn V, Schall JD. (2006) Executive control of countermanding saccades by the supplementary eye field. *Nat Neurosci* **9**: 925.

Stork DG, Wilson HR. (1990) Do Gabor functions provide appropriate descriptions of visual cortical receptive fields? *J Optic Soc Am A* 7(8): 1362–1373.

Styles EA. (2005) Attention, Perception and Memory an Integrated Introduction. Psychology Press, New York.

Swainson R. (2000) ? *Neuropsychologia* **38**: 996.

Swindale NV. (1996) The development of topography in the visual cortex: A review of models. *Comput Neural Systems* 7: 161–247.

Swindale NV, Shoham D, Grinvald A, Bonhoeffer T, Hubener M. (2000) *Nat Neurosci* **3**: 822–826.

Tan SS, Faulkner-Jones B, Breen SJ, Walsh M, Bertram JF, Reese BE. (1995) Cell dispersion patterns in different cortical regions studied with an X-inactivated transgenic marker. *PubMED* **121**(4): 1029–1039.

Tashiro A, Minden A, Yuste R. (2000) Regulation of dentritic spine morphology by the rho family of small GTPases: Antagonistic roles of rac and rho. *Cerebral Cortex* **10**(10): 927–938.

Toal D, Flanagan C, Jones C, Strunz B. (1996) Subsumption architecture for the control of robots. *IMC-13.*

Thomas MSC, McClelland JL. (2008) Connectionist models of cognition. To appear in: Sun R (ed.), Cambridge Handbook on Computational Cognitive Modeling. Cambridge University Press.

Touretzky DS, Saksida LM. (1997) Operant conditioning in skinnerbots. *Adaptive Behav* **5**(3–4): 219–247.

Tournier J, Calamante F, Gadian D, Connely A. (2004) Direct estimation of the fiber orientation density function from diffusion-weighted MRI data using spherical deconvolution. *NeuroImage* **23**: 1179–1185.

Tsang VL, Chen AA, Cho LM, Jadin KD, Sah RL, Delong S, West JL, Bhatia SN. (2006) Fabrication of 3D hepatic tissues by additive photopatterning of cellular hydrogels. *The FASEB Journal, Research Communication, 28 December 2006.*

Tsuda I. (2001) Toward an interpretation of dynamic neural activity in terms of chaotic dynamical systems. *Behav Brain Sci* **24**: 793–847.

Thorpe S, Fize D, Marlot C. (1996) Speed of processing in the human visual system. *Nature* **381**: 520–522.

Trope D, Kreiman G, Lyckman A, Mukherjee S, Yu H, Horng S, Sur M. (2006) Gene expression changes and molecular pathways mediating activity-dependent plasticity in visual cortex. *Nat Neurosci* **9**(5).

Tuch D. (2002) Diffusion MRI of complex tissue structure. *PhD Thesis*, MIT Jan 2002.

Tuch DS. (2004) Q-ball imaging. *Magnetic Reson Med* **52**: 1358–1372.

Tuch DS, Wisco JJ, Khachaturian MH, Ekstrom LB, Kotter R, Vanduffel W. (2005) Q-ball imaging of macaque white matter architecture. *Phil Trans R Soc* **360**: 869–879.

Tulving E. (1972) Episodic and semantic memory. In: Tulving E, Donaldson W (eds), Organization of Memory, pp. 382–402. Academic Press, Inc, New York, NY.

Tulving E, Markowitsch HJ. (1998) Episodic and declarative memory: Role of the hippocampus. *Hippocampus* **8**: 198–204.

Ugurbil K, Toth L, Kim DS. (2003) How accurate is magnetic resonance imaging of brain function? *Trends Neurosci* **26**(2): 108–114.

Ungerleider L, Haxby J. (1994) 'What' and 'Where' in the human brain. *Curr Opin Neurobiol* **4**: 157–165.

Valk PE, Bailey DL, Townsend DW, Maisey MN. (2003) *Positron Emission Tomography: Basic Science and Clinical Practice*. Springer, ISBN 1852334851.

Valiant L. (1994) *Circuits of the Mind*. Oxford Press.

van Gelder TJ. (1998) The dynamical hypothesis in cognitive science. *Behav Brain Sci* **21**: 1–51.

Vargha-Khadem F, *et al.* (1997) Differential effects of early hippocampal pathology on episodic and semantic memory. *Science* **277**: 376–380.

Varshney LR, Sjöström PJ, Chklovskii DB. (2006) Optimal information storage in noisy synapses under resource constraints. *Neuron* **52**(3): 409–423.

Velasquez JD, Maes P. (1997) Cathexis: A Computational Model of Emotions, pp. 518–519. Autonomous Agent, Marina Del Rey, California, USA.

Vincent JL, Snyder AZ, Fox MD, Shannon BJ, Andrews JR, Raichle ME, Buckner RL. (2006) Coherent spontaneous activity identifies a hippocampal-parietal memory network. *J Neurophysiol* **96**: 3517–3531.

Wang RF, Brockmole JR. (2003) Simultaneous spatial updating in nested environments. *Psychonomic Bulletin Rev* **10**(4): 981–986.

Watanabe T, Nanez JE, Sasaki Y. (2001) Perceptual learning without perception. *Nature* **413**: 844.

Weber M, Welling W, Perona P. (2000) Unsupervised learning of models of recognition. *Proc Eur Conf Comp Vis* **2**: 1001–1108.

Wechsler H, Zimmerman GL. (1988) 2-D invariant object recognition using distributed associative memory. *IEEE Transactions on Pattern Analysis and Machine Intelligence*. **10**(6): 811–821.

Wellman HM, Gelman S. (1992) Cognitive development: foundational theories of core domains. *Annu Rev Psychol* **43**: 337–375.

Weng Juyang, Hwang Wey-Shiuan. (2006) From neural network to the brain: Autonomous mental development. *IEEE Computational Intelligence Magazine* **Aug**: 15–31.

Wickelgren Ingrid. (1999) Memory for order found in the motor cortex. *Neurobiol Sci* **283**(5408): 1617–1619.

Williams JHG, Whiten A, Suddendorf T, Perrett DI. (2001) Imitation, mirror neurons and autism. *Neurosci Biobehav Rev* **25**: 287–295.

Willingham DB. (2001) Cognition: The Thinking Animal. Prentice Hall, Upper Saddle River, NJ.

Wilson SW, Ross LS, Parrett T, Easter SS Jr. (1990) The development of a simple scaffold of axon tracts in the brain of the embryonic zebrafish, *Brachydanio rerio. Development* **108**: 121–145.

Wise RA, Rompre PP. (1989) *Annu Rev Psych* **40**: 191.

Wallis G, Oram M. (1997) A model of invariant object recognition in the visual system. *Prog Neurobiol* **51**: 167–194.

Wolfe JM, Horowitz TS, Michod KO. (2007) Is visual attention required for robust picture memory? *Vis Res* **46**(7): 955–964.

Wolfson E, Schwartz EL. (1989) Computing minimal distances on polyhedral surfaces. *IEEE Transactions on Pattern Analysis and Machine Intelligence* **11(9)**.

Yen S-C, Finkel LH. (1998) Extraction of perceptually salient contours by striate cortical networks. *Vis Res* **38(5)**: 719–741.

Zemel R. (2000) Cortical belief networks. In: R Hecht-Neilsen (ed), Theories of the Cerebral Cortex. Springer-Verlag, New York.

Zhang J, Zhang X, Krim H. (1998) Invariant object recognition by shape space analysis. *ICIP 98* **3**: 581–585.

Zola-Morgan SM, Squire LR. (1990) The primate hippocampal formation: Evidence for a time-limited role in memory storage. *Science* **250**: 288–290.

INDEX

A-Brain 167
action potential 24–26, 32
acetylcholine 263, 277
ACT-R architecture 190, 192
adaptive resonance theory (ART) 146
agent 189, 190, 194–199, 203, 204
affective computing 105, 106, 127, 163
affective learning 99, 100, 215
amino acid 263
amygdala 4, 5, 8, 94, 104, 105, 107–110, 113, 115, 226, 249
anterograde amnesia 74, 80
ARTSCAN 49
artificial intelligence (AI) 35, 120, 128, 165, 166, 174–176, 186, 187, 199, 217, 218, 220, 258
artificial neural networks (ANN) 20, 34, 166, 245
artistic 170, 213
association cortex 86, 87, 97, 178, 252
associative learning 95
associative network 223
attention mechanism 251
autistic 106, 117, 125, 258
axon 2, 3, 14, 21, 24, 26, 36, 43, 44, 111, 130–133, 138, 140, 149, 170, 184, 225, 227, 236
axon terminal 24, 26
augmentation 252, 253
autonomic nervous system 264

background emotions 106
basal ganglia 5, 8, 42, 71, 74, 96, 99, 193, 251
Bayesian mind 153
Bayesian networks 52, 154, 159, 160, 162
B-Brain 167
BDI architecture 187, 196–199, 204
biologically-inspired cognitive architecture (BICA) 208
Blue Brain 16, 183, 184
BOLD fMRI 234
brain lesions 113, 241
brain-inspired computing 221, 259
brain rhythms 319, 324

brain stem 1, 3, 5, 8, 83, 111
brainwaves 230

CAI, CA2, CA3 156
Cartesian dualism 257
cathexis 100, 124, 125
C-Brain 167
central nervous system (CNS) 19
cerebellum 3, 8, 69, 71, 74, 96, 192, 251
cerebral cortex 1, 3, 5, 8, 42, 44, 73, 92, 129–131, 151, 192, 251
chloride (Cl-) 25
cingulate gyrus 110, 111
circumplex model 119
clarion architecture 201, 202
classical conditioning 71, 91, 92
cognitive 3, 5, 16, 35, 44, 46, 47, 49, 72, 74, 80–82, 85, 96, 98–101, 105, 106, 112, 113, 115, 117, 120, 121, 123, 124, 126–129, 133, 137, 146, 151, 153–155, 157–160, 162, 166, 172, 173, 182–189, 191, 193, 194, 197, 199, 200, 202, 204–209, 217, 219, 221, 230, 231, 236, 241, 243, 248, 249, 257, 259
cognition 90, 96, 99, 104–106, 120, 126, 127, 129, 133, 142, 145, 151, 152, 154, 155, 164, 168, 173, 182, 183, 185, 187, 191–193, 204, 209, 243, 247, 254
cognitive learning 82, 99
cognitive neuroscience 235
cognitive psychology 154, 186, 187 191
cognitive science 47, 106, 124, 221
cognitive system 155, 221
cognitive computing 183, 184
commissure 266
commonsense knowledge 81, 163, 165, 173–176, 181
complementary 45, 46, 60, 84, 142–144, 183, 248
complex cell 52, 53, 248
competitive 45, 46, 60, 138, 144, 145, 147, 247, 248
computational neuroscience 266
computational power 2, 13–15, 145, 184, 244, 246, 253, 254
computational theory of mind 259
conceptual blending theory 164, 172, 181
conditional stimulus (CS) 71, 92, 94, 98
conditional learning 90, 92, 95, 99
corpus callosum 1, 3, 5, 110
cortex 1, 3–9, 11–13, 21–23, 27–39, 33–49, 51, 56, 58–61, 66, 69–75, 77, 86–88, 90, 92, 94, 97, 98, 101, 107, 112–116, 129–134, 136, 137, 139, 140, 142, 147, 149, 151, 153, 178, 183, 184, 192, 217, 223, 225, 226, 228, 235, 245, 247–249, 251, 252
  cerebral cortex 1, 3, 5, 8, 42, 44, 73, 92, 129–131, 151, 192, 251
  cortex architecture 35, 37, 38, 43, 44, 46, 60, 153, 245
creative intelligence 170, 171
critical period 37, 83, 85, 86
CT scan 240

current 15, 21, 24, 33, 34, 46, 48, 51, 74, 94, 100, 140, 145, 157, 160–162, 165, 184, 185, 188, 189, 195, 203, 205, 208, 216, 217, 221, 224, 232, 247, 253, 254, 256
Cyc 165, 174, 175, 181
cytoarchitecture 229

DARPA 34, 129, 184, 208, 254
D'Brain 155, 156
declarative memory 7, 67–72, 74, 79, 80, 159, 207
decision making 5, 38, 58, 59, 61, 65, 95, 96, 100, 104, 105, 115, 116, 126, 128, 178, 188, 198, 207, 226
dendrites 21, 24, 36, 130–133, 140
diffusion 23, 235, 236, 237
  diffusion MRI 233, 235, 236
distributed 125, 201, 207, 223, 246, 251
  distributed representation theory 29
  distributed neural system 246
dopamine 32, 93, 94, 98, 99, 111, 112, 247
dual augmentation strategy 252, 253
dynamic pointers 45
dynamic reasoning 155, 161, 162
dynamical system 224

electroencephalography (EEG) 230, 231–234, 237, 238, 240
emotion 8, 99, 100, 104–108, 110, 115–128, 16, 164, 168, 178, 180, 208, 221, 226, 256
  background 106, 124
  primary 106–108, 117, 119, 120, 123
  secondary 106, 107, 123
emotion machine 163
emotional system 128
emotional state 76, 106, 115–117, 124, 215
episodic memory 69, 70, 72, 80, 155, 157, 159, 189, 190, 205
event related potentials (ERP) 230–234
excitatory 8, 25, 26, 33, 35, 46, 49, 51, 52, 94, 138, 140, 147, 149, 232, 245
  excitatory neuron 245
  excitatory neurotransmitters 26, 267

feedforward and feedback 40, 44, 138, 147
folded feedback 140, 141
fragment 160–162
free recall 151
frontal lobe 3, 60, 84, 116
functional magnetic resonance imaging (fMRI) 59, 91, 228, 230, 233–236, 240, 241
fusiform face area 9, 10, 56, 57, 247, 248, 251

GABA (gamma-aminobutyric acid) 26
ganglion cell 14, 52
gene 178, 246
general intelligent systems 221
general artificial intelligence 221
glutamate 94

gray-matter 84
grandmother cell theory 10, 28

hierarchical 29, 38–44, 46–48, 52–58, 60, 153, 193, 247
hierarchical pathways 43
hierarchical temporal memories (HTMs) 52, 53
hippocampus 4, 5, 7, 62, 67–75, 80, 83, 97, 107–110, 115, 127, 155, 159, 193, 226, 251
Hodgkin-Huxley's cell membrane equation 184
horizontal interaction 35, 44, 48, 49, 60, 131, 135, 136, 147
human intelligence 1, 164, 165, 170, 186, 244, 252, 253, 256
human learning 82, 86, 87, 98, 99, 102, 247
human memory 62, 75–78
Huntington's disease 8
hypothalamus 5, 8, 94, 107, 110

IBM 16, 183, 214
ICARUS architecture 193
immune system 118
inferior temporal cortex (IT) 87, 113, 130, 142, 153
information theory XXX
information processing system 225, 226
inhibitory 8, 25, 26, 35, 46, 49, 51, 52, 91, 139, 140, 147, 149, 231, 244
integrated cognitive architecture 185, 186
intent inference 157, 158
ion 25, 26, 184
Ivan Pavlov 92

knowledge machine 175
knowledge representation 159, 173, 174, 191

laminar computing 129, 131, 133, 134, 136, 140, 144, 147
laminar pattern 130, 131
LAMINART model 146–150
learning 13, 18, 20, 31, 35, 37, 44, 46–48, 50–53, 60, 62, 69, 70, 81–103, 123, 125, 126, 129, 136–138, 141, 142, 144–147, 151, 152, 155–159, 173, 183, 187, 190, 195, 198, 199, 201–204, 207, 212, 215, 223, 247, 250, 251
lesions method 237, 240, 241
limbic system 1, 3, 5, 104, 108–110, 112, 115, 116, 155
long-term depression (LTD) 31, 32, 36, 67, 82–84, 244
long-term memory (LTM) 31, 56, 63, 66–70, 74, 79, 81, 157, 158, 160, 162, 183, 188, 189, 192–195, 208
long-term potentiation (LTP) 31, 32, 36, 97, 82–84, 244

map machine 224
magnetic field 232, 234–237
magnetic resonance imaging (MRI) 59, 232, 234, 235, 239
magnetoencephalography (MEG) 236, 237, 239, 240
match and mismatch learning 101

memory systems 45, 60, 62, 63, 67, 74–76, 79, 80, 102, 208, 253
mental state 117, 256–259
mental models 156–161
mind chip 318
mind-brain identity 256, 257
mirror neurons 97, 246
molecule 26, 182, 184, 246
Moore's law 15, 219
motor learning 86, 95, 96, 250
multi-agent system 199
multi-modality 247
multiple intelligences 164, 168, 180, 224
myelin 235

near-infrared spectroscopy 239
neocortex 1, 13, 16, 44, 100, 107, 115, 130–133, 155, 157
neural 10, 13, 19–21, 27, 34, 35, 37, 49, 56, 72, 77, 81–83, 87, 88, 95, 102, 107, 111, 113, 124, 125, 129, 133, 137, 138, 144, 146, 149, 151, 153, 159, 166, 187, 201–203, 208, 224–226, 229, 230, 233, 234, 240, 245, 246, 248–250, 252
  neural plasticity 82, 83, 102, 250
  neural cortical column 13, 35
neurogenesis 84
neurophysiology 259
neuroscience 1, 38, 74, 222, 229, 235
neuron (nerve cells) 1–3, 9, 10, 14–16, 18–36, 38, 46, 52, 56, 57, 67, 72, 75, 77, 82, 83–5, 94, 95, 97–100, 111, 112, 130–133, 136–140, 149, 166, 183–185, 213, 223–227, 230–234, 238, 240, 244–248, 251, 252, 258
neuron population encoding 22
neural networks 20, 34, 35, 81, 113, 129, 144, 159, 166, 201, 203, 208, 226, 246, 248, 249
neurotransmitters/neuromodulators 21, 26, 32, 124, 231, 246
non-declarative memory 67–71, 74, 79
nucleus 5, 6, 21, 31, 71, 94, 109–112, 142, 147, 247

occipital lobe 3, 4, 192
ontology 174, 175, 216
operant conditioning 92, 93
open-ended-open minded design 178, 252
open mind 165, 174, 175, 251
optic nerve 6, 7, 14, 29
optical imaging 238
orbitofrontal cortex 113
organismal level 183, 243, 248
Ortony Clore Collins cognitive model (OCC) 100, 120, 121, 123

Papez's circuit 107, 108
parallelism 244, 245
parahippocampal place area 9, 10
parietal lobe 3, 4, 84
Pavlovian 92, 94
Parkinson's disease 8, 98
pattern recognition 137, 146, 178, 193, 249
peptide 223
perception 4, 6, 9, 23, 24, 53, 77, 104, 105, 130, 136, 145, 157, 158, 188, 189, 193, 196, 197,

198, 200, 204, 205,. 207, 216, 256
perceptual learning 53, 86–88, 90, 91, 95, 102, 137, 142
phonological loop 65
pinwheel 132, 133
plasticity 12, 13, 20, 31, 32, 82–86, 101, 102, 137, 138, 145, 166, 180, 184, 249–251
positron emission tomography (PET) 225, 238–240
postsynaptic neuron 24–26, 32, 84
potassium (K+) 24
practical intelligence 171
prefrontal cortex 5, 38, 41, 46, 48, 51, 56, 58–61, 66, 94, 98, 112, 113, 115, 142, 151, 192
probabilistic computing 153
probabilistic model 155, 162
procedural memory 67, 70, 71, 159, 188
primary emotions 106, 117, 119, 120, 123
primary motor area 272
primary sensory area 272
  primary visual cortex 4–6, 30, 288
  primary auditory cortex 4
  primary motor cortex 5, 27, 28
psychophysics 272
Purkinje cell 272
pyramidal cell 131, 136, 140, 149

quantum machine XXX

receptor 21, 24–26, 30, 70, 71, 83, 84, 100, 112, 225, 247
receptive field 20, 29–31, 48, 49, 248
regionalization 2, 10, 37
reinforcement learning 93–95, 99, 190, 202, 203, 207, 247, 251
relational learning 82, 85, 86, 97
retina 6, 7, 14, 24, 29, 30, 58, 61, 225
retrograde amnesia 281
retrosplenial cortex 9
reverse hierarchy theory 53, 54
Roseman's cognitive appraisal model 121
Russell's model 119

saccade 42, 43
second messenger 31, 247
secondary emotions 106, 123
self-reflective system 167
semantic memory 67–70, 157–159, 189, 190, 207
sensory memory 63, 64, 79, 90
septum 112
short-term memory 56, 63, 66, 67, 72, 74, 79, 98, 138, 145, 157, 158, 160, 194, 208
shunting networks 138, 145, 185, 224, 245
simple cell 48, 52, 248
singularity 273
SOAR architecture 187, 188, 190
society of mind 118
soma 21
sodium (Na+) 25
somatosensory 4, 5, 114, 138
spike patterns 32, 33
spiking 32, 33, 35, 36, 74, 184, 185, 245
  spiking model 35, 36, 245

split-brain 5
specialization 2, 10, 37
specialized neurons 21, 46
stability 48, 82–84, 101, 102, 145, 166, 249
stability-plasticity dilemma 101, 145, 249
stimulus-response learning 86, 92, 94, 95
strong artificial intelligence (Strong AI) 165
substantial nigra 274
subsumption architecture 187, 199–201, 204
sulcus 110
superior colliculus (SC) 9, 42, 51
supervised learning 51, 60, 95, 100, 101, 137, 146
SyNAPSE project 254
synaptic vesicle 274
synaptic potential or synaptic strength 25–27, 31, 32, 36, 67, 245
synaptic plasticity 20, 31, 32, 251
synaptogenesis 67
system neuroscience 275

temporal difference learning 93–95, 99
temporal lobe 3, 4, 7, 67–69, 71, 72
temporal resolution 88, 231, 233, 238, 240
thalamus 3, 5, 8, 9, 107, 109, 110, 112, 114, 155, 157, 193, 226, 249
theory of mind 257–259
theory of theory 257
transcranial magnetic stimulation (TMS) 238, 240
three-layer architecture 123, 124
triarchic theory of intelligence 164, 170, 181
Turing test 164, 213, 214, 217–219

uncertainty principle 274
unsupervised learning 48, 51, 60, 99–101

ventral tegmental area (VTA) 94, 111
vision 11, 29, 44, 49, 53, 85, 129, 132, 133, 142, 143, 145, 146, 151, 152, 180, 192, 213, 215, 216
visual cortex 4–7, 11, 12, 21–23, 29, 30, 38–42, 44, 46, 47, 49, 60, 87, 88, 90, 101, 129, 130, 132, 133, 142, 147, 149, 178, 225
visuospatial sketchpad 65, 66

white matter 74, 80, 131, 235
wisdom 165, 176–181
wordnet 174, 175
world wide web 223
working memory 46, 56, 58, 63–66, 79, 115, 117, 151, 152, 160, 162, 172, 183 188, 189, 192, 204–207